Collins

Maths Frameworking

3rd edition

Rob Ellis, Kevin Evans,
Keith Gordon, Chris Pearce,
Trevor Senior, Brian Speed,
Sandra Wharton

William Collins' dream of knowledge for all began with the publication of his first book in 1819. A self-educated mill worker, he not only enriched millions of lives, but also founded a flourishing publishing house. Today, staying true to this spirit, Collins books are packed with inspiration, innovation and practical expertise. They place you at the centre of a world of possibility and give you exactly what you need to explore it.

Collins. Freedom to teach.

Published by Collins
An imprint of HarperCollins*Publishers*
77–85 Fulham Palace Road
Hammersmith
London
W6 8JB

Browse the complete Collins catalogue at
www.collins.co.uk

10 9 8 7 6 5 4 3

ISBN-13 978-0-00-753787-7

British Library Cataloguing in Publication Data
A Catalogue record for this publication is available from the British Library.

Written by Rob Ellis, Kevin Evans, Keith Gordon, Chris Pearce, Trevor Senior, Brian Speed and Sandra Wharton
Commissioned by Katie Sergeant
Project managed by Elektra Media Ltd
Development edited and copy-edited by Gudrun Kaiser
Edited by Helen Marsden
Proofread by Grace Glendinning
Illustrations by Jerry Fowler and Ann Paganuzzi
Typeset by Elektra Media Ltd
Cover design by Angela English
Production by Rachel Weaver

Acknowledgements
The publishers wish to thank the following for permission to reproduce photographs. Every effort has been made to trace copyright holders and to obtain their permission for the use of copyright materials. The publishers will gladly receive any information enabling them to rectify any error or omission at the first opportunity.

Cover Hupeng/Dreamstime

Contents

8 Algebra

9 Decimal numbers

10 Surface area and volume of 3D shapes

11 Solving equations graphically

12 Distance, speed and time

13 Similar triangles

14 Revision and GCSE preparation

Introduction

Maths Frameworking has been fully revised and updated in line with the 2014 Programme of study for Key Stage 3 Mathematics. This third edition provides complete coverage of the subject content of the new curriculum, as well as ample material to support pupils in fulfilling its three overarching aims: developing mathematical fluency, mathematical reasoning, and problem-solving skills.

Maths Frameworking 3rd edition components

- **Pupil Books:** three per year, catering for different levels of ability.
- **Teacher Packs:** one to accompany each Pupil Book.
- **Homework Books:** one per year, encompassing practice material that is graduated in difficulty and suitable for all ability levels.
- **Intervention Workbooks:** five workbooks of increasing difficulty level (levels 3–8 of the previous national curriculum), designed for targeted one-to-one intervention to help pupils achieve the expected level of progress.
- **Digital resources on Collins Connect:** Collins Connect is an online platform, which includes interactive versions of the Pupil Books, with interactive classroom and homework activities, assessments, tools, and videos that have been specially developed to help pupils improve their mathematics skills.

Features of this Teacher Pack

This Teacher Pack accompanies *Maths Frameworking 3rd edition Pupil Book 3.1* and caters for lower-tier pupils (working at roughly Levels 5–6 of the previous curriculum). Medium-tier pupils (working at roughly Levels 5–7) are catered for by Pupil Book 3.2 and the accompanying Teacher Pack 3.2, and higher-tier pupils (Levels 6–8, with a greater proportion of level 7 and 8 materials) are catered for by Pupil Book 3.3 and Teacher Pack 3.3. The topics and their sequence are the same across all three books but are handled at different levels, allowing for sensitive differentiation in mixed-ability classes.
We have produced around 100 one-hour lessons to cover all the material you need in order to deliver the new Programme of study. These lessons should provide the flexibility to include tests, extended activities and revision classes in the teaching programme. The lessons also allow for the normal events that may disrupt teaching time. For further help when mapping lessons and producing your scheme of work, please consult the Contents pages and the Programme of study matching chart provided in this book.

Chapter overview

Pupil Book chapters are categorised according to the sections of the Programme of study to which they relate, and are colour-coded as follows:

- Number – pink
- Algebra – green
- Ratio, proportion and rates of change – lilac
- Geometry and measures – orange
- Probability – navy
- Statistics – teal

Each chapter of this Teacher Pack starts with an outline of the content covered in the chapter, explaining clearly how the lesson plans cover the National Curriculum and providing context for the topics covered. Its features are listed below.

- **Learning objectives** identify the key learning outcomes.
- **Prior knowledge** highlights the underpinning mathematics that pupils will draw on in the chapter, taking into account the coverage of the new Programme of study for Key Stage 2.
- **Context** provides some real-life or historical background for the key mathematical ideas explored in the chapter and gives suggestions on how to use the corresponding chapter opener in the Pupil Book.
- **Discussion points** provide some ideas for warm-up questions to introduce pupils to the topic.
- **Associated Collins ICT resources** provide an overview of the digital resources related to the chapter, which are available on the Collins Connect online platform.
- **Curriculum references** show how the materials meet the requirements of the new curriculum with references to the 2014 Programme of study.
- **Fast-track for classes following a 2-year scheme of work** provides suggestions for parts of the chapter that can be omitted if the class is following a condensed Key Stage 3.

Lesson plans

A lesson plan is provided for each topic in the Pupil Book. Every plan follows the same format, making it easy to prepare for and use. Its features are listed below.

- **Learning objectives** identify the key learning outcomes of the lesson.
- **Links to other subjects** highlight the topics' cross-curricular links, encouraging pupils to relate what they are learning in mathematics to what they are learning in other subjects.
- **Resources** list the content relating to the lesson that is available in other components of the scheme.
- **Homework activities** provide references to the related activities in the Homework Book and on the Collins Connect online platform.
- **Key words** provide a list of the key terms that pupils need to understand and use to talk about this topic. The words match the key words in the Pupil Book.
- **Problem solving, reasoning and financial skills help** details how the topic relates to the overarching aims of the Programme of study and, where appropriate, how it helps pupils to develop their financial skills.
- **Common misconceptions and remediation** pinpoints typical problems that pupils may have in getting to grips with the topic, and gives advice on how pupils can overcome these problems.
- **Probing questions** are designed to bring out important aspects of the topic and to link these aspects to pupils' prior knowledge.
- **Part 1** of the lesson plan is an engaging oral and mental starter involving the whole class; it has been designed to require minimal specialised equipment.
- **Part 2** is a main lesson activity, which helps you to lead pupils into the exercise questions.
- **Part 3** is a plenary designed to round off the three-part lesson.
- **Answers** are provided at the end of each lesson plan. Answers for the review questions and the activity are at the end of each chapter.

Answers for other course components

Answers to the Homework Book and Intervention Workbook questions can be found at this website link: **www.collins.co.uk/mathsframeworkinganswers**.

Scheme of work

A flexible 2-year and 3-year scheme of work is provided at the back of the book and in editable Word and Excel format on the CD-ROM accompanying this book. It shows two routes through the content, which are suitable for schools following a 2- or 3-year programme for Key Stage 3.

Teacher Pack CD-ROM

The CD-ROM in the front cover of this book contains:

- all the lesson plans in Word format for you to customise as you wish
- the 2- and 3-year schemes of work in Word and Excel format
- a printable version of the progression checklists for each chapter of the Pupil Book, which may be copied and distributed to the class for self-assessment, and for pupils to gauge their progress and learn how to improve
- skills maps for fluency, mathematical reasoning and problem solving
- a letter describing the philosophy and features of *Maths Frameworking 3rd edition* that can be customised and distributed to parents.

Digital resources available on Collins Connect

A variety of engaging resources are available on Collins Connect to support and enhance your teaching.

- **Classroom presentation resources** include:
  - an interactive version of the three Pupil Books for each year
  - interactive activities, including matching pairs and drag-and-drops, that can be used as starters and plenaries to improve fluency or stimulate group discussion
  - engaging video clips that develop pupils' conceptual understanding of key topics
  - 'Maths Man' videos that use rhythm and music to help pupils remember essential maths facts
  - exciting real-life videos and images, with accompanying teacher notes, that bring to life the awe and wonder of mathematics and encourage open conversations involving mathematics
  - audio-worked examples that model how to approach a variety of question types – including problem solving – and demonstrate best practice.
- Innovative **skills-building interactives** that enable pupils to explore and discover confidentially in order to build their conceptual understanding of, and skills in proportional reasoning and equations.
- A **digital version of the Pupil Book glossary**, complete with recordings of all the key words – ideal for supporting literacy.
- **Auto-marked homework** tasks to accompany each lesson. Differentiated homeworks are provided for in the different Pupil Books.
- **Auto-marked assessments with diagnostic feedback**, including a diagnostic assessment for the start of the course, half-term assessments covering the main

chapter content, and an end-of-year test, which is split into two halves so that it fits easily into lesson time.

Collins Connect also provides invaluable tools to aid in tracking pupils' progress:
- **Synchronisation with SIMS**, enabling you to set up individual pupil accounts quickly and easily.
- **Task assignment** functionality, which lets you set any part of the interactive Pupil Book to a pupil or class as an assignment. As with the interactive homework tasks, you can set pupils a deadline for completion.
- **Results overviews** show the marks achieved by your classes in the homework tasks and tests you have set for them. Results can be displayed by a pupil, an assignment or a class, so that you can pinpoint areas for intervention easily.

Find out more at **www.collins.co.uk/connect**.

Maths Frameworking and the 2014 Key Stage 3 Programme of study for mathematics

Rationale

Following the publication of the Schools White Paper, 'The Importance of Teaching' (DfE, November 2010), the UK government instigated a review into the national curriculum in mathematics. The stated aim of the review was to benchmark expectations for pupils in this country against the expectations of the most successful nations in the world, and to draw on this information to design a new curriculum that would put English pupils on a par with pupils in the highest-performing countries. In February 2013, the DfE began an extensive consultation process involving the mathematics community. In September 2013, this process resulted in the publication of a revised Programme of study for Key Stage 3 mathematics. The new curriculum became statutory in September 2014.

The 2014 Programme of study for mathematics

Subject content

The subject content of both the new Key Stage 2 and Key Stage 3 programmes of study has changed with respect to the former curriculum. Probability has been removed from the Key Stage 2 curriculum, in favour of greater emphasis on the foundations of algebra and on skills in number (for example, multiplication tables up to 12×12 and long multiplication and division). This means that the expectations of the prior knowledge that pupils bring with them at the start of Key Stage 3 will also change. Although there has been some reduction of content, in other areas the expectation in Key Stage 2 is very demanding as it includes some topics that were at level 6 in the old Programme of study. This third edition of *Maths Frameworking* considers these new expectations and creates a framework for progression from this new starting point.

At Key Stage 3, the Programme of study is divided into six strands. Within these, some of the content requirements have been more fully described. One example is a deeper exploration of the properties of prime numbers than before. There is also increased precision in the areas of algebra and of geometry and measures. 'Ratio, proportion and rates of change' now stands alone as a strand in its own right, emphasising the links between fractions, ratios, proportion and percentages. Probability and Statistics are now also treated as separate strands. Some of the material that was at level 8 in the old Programme of study has been removed. Trigonometry is an example. The intention is that pupils who grasp concepts rapidly should be challenged with more demanding problems on the same subject matter rather than being accelerated through new content in preparation for GCSE. The structure and coverage of the *Maths Frameworking* Pupil Books has been reworked comprehensively to reflect these changes.

Overarching aims

A key difference in the new Programme of study for mathematics is the introduction of three overarching aims, which are presented as integral to study of the subject content. The aims are to ensure that all pupils:

- become fluent in the fundamentals of mathematics, including through varied and frequent practice with increasingly complex problems over time, so that pupils develop conceptual understanding and the ability to recall and apply knowledge rapidly and accurately.
- reason mathematically by following a line of enquiry, conjecturing relationships and generalisations, and developing an argument, justification or proof using mathematical language.
- can solve problems by applying their mathematics to a variety of routine and non-routine problems with increasing sophistication, including breaking down problems into a series of simpler steps and persevering in seeking solutions.

The Key Stage 2 and Key Stage 3 programmes of study share the same overarching aims, and these aims form the basis for the assessment objectives in the new mathematics GCSE for first teaching in September 2015. The intention of the DfE is to ensure smooth progression through the three Key Stages, with the Key Stage 3 Programme of study consolidating and building on understanding from Key Stage 2 in order to form solid foundations for further study at GCSE and beyond.

The aims of the national curriculum are intended to be developed and applied across the mathematical content of the Programme of study, and their treatment in *Maths Frameworking 3rd edition* reflects this. For example, pupils develop fluency by using algebra to consolidate their understanding from Key Stage 2 and generalise the structure of arithmetic. They reason mathematically and extend their understanding of the number system by making connections between number relationships and their algebraic and graphical representations. They solve problems by selecting appropriate concepts, methods and techniques to apply to unfamiliar and non-routine problems.

Fluency and linking

Fluency in mathematics has two important facets: familiarity and recollection. Familiarity is important because it means that, when faced with a calculation, pupils can recognise the steps required to complete it by drawing on prior experience of completing other similar calculations. The ability to recollect pertinent mathematics facts and techniques further enhances pupils' ability to complete calculations rapidly, as they access, for example, their knowledge of times tables and mental mathematics techniques to reach an answer without needing to use a calculator.

Fundamental to fostering both of these aspects of mathematical fluency, is ensuring that pupils gain extensive practice. Rich and varied practice lies at the heart of *Maths*

Frameworking. At the start of each lesson plan in this Teacher Pack, we suggest a series of probing questions, which are designed to encourage pupils to recall previously learnt mathematics facts and skills that are of relevance to the topic. The exercises within every topic offer a wide range of questions of increasing complexity to which pupils can apply these skills and knowledge. The Homework Books and auto-marked homework activities on Collins Connect offer further opportunities for practice.

In addition to ensuring plentiful practice, the *Maths Frameworking* scheme ensures that the practice encourages in pupils the understanding and mastery that are the hallmarks of true fluency by providing opportunities for practice across different mathematical and cross-curricular contexts.

In Ofsted's paper, 'Mathematics: Understanding the Score' (Ofsted, September 2008), Ofsted identified that the majority of pupils had too few opportunities to use and apply mathematics and to make connections across different areas of the subject. In the 2014 national curriculum, mathematical topics figure in a wide variety of curricular subjects. For example, Design and technology pupils need to be comfortable when working with 2D and 3D plans, while data-handling skills are required in Geography, Science and Computing. The 2014 national curriculum also places renewed emphasis on financial education, with a focus within the new mathematics Programme of study on solving problems involving percentage increases and decreases, simple interest and repeated growth. This has clear links to the Programme of study for Citizenship, which requires pupils to learn the functions and uses of money, the importance and practice of budgeting, and managing risk.

Maths Frameworking draws out these intra- and cross-curricular links in order to help pupils to become adept at making connections. Throughout the books, questions are presented in varied contexts taken both from real life and from the subject areas mentioned previously, plus other subject areas. In the 'Review' questions at the end of each chapter, as well as checking pupils' grasp of the mathematics topics covered within it, other parts of mathematics are also brought into play. Questions are posed, which require pupils to combine learning across different topics in order to answer them. Furthermore, each chapter begins with an opening page that presents real-life applications of the topic covered, giving pupils a window onto why it is important and useful outside of the classroom. This helps pupils to make links between the mathematics they are studying in different parts of the curriculum, to the relevance in real life, rather than learning only the mathematics. These Pupil Book features are supported by accompanying guidance in this Teacher Pack: the 'Context' section at the start of each chapter provides background information for teachers, while the 'Links to other subjects' section of each lesson plan highlights cross-curricular links in each topic.

Problem solving and reasoning

The ability to solve problems within mathematics and in other subjects is a key skill, which consultations have shown that employers require across different industries and economic sectors. In the specifications for the new mathematics GCSE for first teaching in September 2015, problem solving is to become one of the three assessment objectives, together with using and applying standard techniques and reasoning mathematically. Teaching at Key Stage 3 needs to build pupils' problem-solving skills in preparation for study at GCSE level. Crucial to this is equipping pupils with strategies that will help them to see the 'big picture' and nurture mathematical independence, allowing time for thinking, and encouraging discussion in the classroom. The lesson plans in this book provide a wide variety of suggestions on how teachers can do this, and offer varied opportunities for probing pupils' understanding and encouraging reflection.

The first step in problem solving in mathematics at this level is often identifying 'where the maths is' and what mathematical techniques can be applied to answer a question. The problem-solving questions in *Maths Frameworking*, especially the extended double-page problem-solving spreads, guide pupils through this process, equipping them with techniques to decipher word problems and 'translate' descriptions into mathematics. Making conjectures and trying different approaches to solve problems is also an important reflex that *Maths Frameworking* develops by providing guided questions in which pupils can experiment and are encouraged to question and reflect. Across the varied problem-solving questions in *Maths Frameworking*, pupils are also encouraged to develop visual representation techniques by using graphs, charts or diagrams to help them get to grips with a word problem or identify a pattern.

Mathematical reasoning is another key skill that has been embedded in this third edition of *Maths Frameworking*. Pupils are reasoning mathematically when they follow a line of enquiry, conjecture a relationship or generalisation, and develop an argument, justification or proof using mathematical language. Reasoning questions and extended double-page reasoning spreads, give pupils opportunities to do this, encouraging them to develop the skill set needed in order to investigate, prove and explain. Developing good habits for showing working and justifying answers will stand pupils in good stead for the second new assessment objective at GCSE level, for which they will need to demonstrate their ability to: make deductions, inferences and draw conclusions from mathematical information, construct chains of reasoning to achieve a given result, interpret and communicate information accurately, present arguments and proofs and assess the validity of an argument and critically evaluate a given way of presenting information. Giving pupils the opportunity to improve the quality of their reasoning is also important in developing their abilities in this area. The 'Common misconceptions and remediation' section of each lesson plan aims to guide teachers in helping pupils to improve.
Providing opportunities for extended responses builds mathematical resilience in pupils, and is a necessary skill for problem solving and reasoning. In the Pupil Book, particular types of question are flagged so that teachers and pupils can find them easily and see which skills are being worked on:

indicates questions that require pupils to solve problems

indicates questions that require pupils to reason mathematically

indicates questions that help to develop pupils' financial skills and financial awareness.

Approach to progression

In the 2014 curriculum, the system of levels formerly used to report pupils' attainment and progress has been removed. The reason given for this by the DfE was that the over-emphasis on levelling in schools encouraged teachers to focus too tightly on 'hitting the numbers' rather than forming a more holistic view of each pupil's progress in terms of conceptual understanding. Ofsted subject-specific exemplification states that the 'development of all pupils' conceptual understanding [...] and progression within each lesson and over time' is one of the hallmarks of outstanding mathematical teaching. In keeping with this, the new Programme of study sets out what subject knowledge should be acquired by the end of Key Stage 3, but gives schools the freedom to develop their own scheme of work for covering it, and their own system for tracking pupils' progress.

The scheme of work and lesson plans in *Maths Frameworking* offer schools a framework structured to ensure progression both over time and within every lesson. The scheme of work has been developed to ensure that all pupils meet expectations in relation to the aims and content of the Key Stage 3 Programme of study, and that they have the opportunity to progress through the curriculum in a way that supports understanding and challenge. We appreciate that giving pupils the *appropriate degree* of challenge for their current level of ability and knowledge is crucial to their making progress, and the three different tiers of the *Maths Frameworking* scheme enable us to achieve this by creating differentiated progression pathways for pupils according to their different starting points. Pupil Book 3.1 covers levels 4–6 of the former national curriculum; Pupil Book 3.2 covers levels 5–7, and Pupil Book 3.3 covers levels 6–8, with a greater proportion of levels 7 and 8 materials. We have consulted with mathematics teachers and experts on how material should be presented and structured within the different tiers in order to ensure smooth progression for different pupils. The result is materials that are tailored to pupils' differential prior knowledge, skills and understanding, which gradually build on these to foster and support progress.

Progression is built into every exercise in the scheme. Exercises start with straightforward questions to consolidate skills and understanding; then move on to more varied and demanding material, and end with extension tasks that are designed to challenge and stretch pupils at the level that is appropriate to their learning abilities. At the end of each chapter of the Pupil Book, pupils are encouraged to reflect on what they have learned and what they need to do to progress by self-assessing using the 'Ready to progress?' checklist (also available in printable format on the CD-ROM accompanying this book).

To enable schools and teachers to check that pupils are on track to meet expectations, we have developed a system of colour-coded icons to show the level of difficulty of each question in the Pupil Books. Teachers and pupils can see at a glance whether they are working in line with expected progress for each learning pathway and year of study. We appreciate that, during the first stages of implementing the new curriculum, many schools will continue to monitor pupils' progression in terms of the levels used in the former curriculum. To facilitate this, and aid in the transition between this system and the new system, which is based on expectations of progress, we have based our system of icons on the former level descriptors. The table on the next page shows the approximate correlation between the progress indicators used in *Maths Frameworking* and the levels used in the previous national curriculum:

	Less than expected progress	Expected progress	More than expected progress
Pupil Book 1.1	≤ Level 3	Level 4	≥ Level 5
Pupil Book 1.2	≤ Level 4	Level 5	≥ Level 6
Pupil Book 1.3	≤ Level 4	Level 5	≥ Level 6
Pupil Book 2.1	≤ Level 4	Level 5	≥ Level 6
Pupil Book 2.2	≤ Level 5	Level 6	≥ Level 7
Pupil Book 2.3	≤ Level 5	Level 6	≥ Level 7
Pupil Book 3.1	≤ Level 4	Level 5	≥ Level 6
Pupil Book 3.2	≤ Level 5	Level 6	≥ Level 7
Pupil Book 3.3	≤ Level 6	Level 7	≥ Level 8

Another aspect of progression that has been emphasised in recent studies on mathematics pedagogy is the CPA, or concrete-pictorial-abstract, sequence. Research suggests that progression in mathematical thinking rests on transitioning from the physical manipulation of concrete materials (an important feature of practice at primary level) to pictorial representations, and finally to written representations in mathematical notation. Developing connections between these three types of experiences, and thereby transitioning through the three stages of understanding, is foremost for making solid progress and becoming mathematically confident and fluent. *Maths Frameworking* aids pupils in building on their experiences in Key Stage 2 and progressing to pictorial and abstract representations by drawing on a variety of different media, for example, pictures, diagrams and charts, as well as correct mathematical notation, in order to explain concepts. Where appropriate, for concepts that are being introduced for the first time, the lesson plans in this Teacher Pack provide suggestions for guidance to use in class.

Approach to assessment

In its paper, 'Mathematics: Understanding the Score', Ofsted identified the need for teachers to place greater emphasis on developing pupils' understanding and on checking it throughout lessons. Ofsted advocated that schools should ensure that pupils have a wide range of opportunities to use and apply mathematics, and that these should be underpinned by thorough assessment, recording and reporting. Regular assessment is known to be a motivating factor for pupils, as gaining awareness of the progress they are making helps to build their confidence, thereby laying the foundations for further achievement. The abolition of levels in the 2014 national curriculum will see a phase shift in how schools approach assessment; *Maths Frameworking* offers a range of tools to support schools in making this transition.

Under the new curriculum, it is the intention of the DfE that assessment should be built into and become integral to the school curriculum, thus allowing schools to check what pupils have learned and whether they are on track to meet expectations at the end of the Key Stage ('Assessing Without Levels', DfE, June 2013). This means creating opportunities for formal, periodic assessment and building informal assessment into daily classroom practices through, for example, the effective use of probing questions to assess progress and identify and tackle misconceptions.

Maths Frameworking supports schools' own approaches to formative assessment by providing a suite of resources that will enable them to track and offer evidence of each pupil's progress through the Key Stage, however and whenever they desire.

- Discussion points are suggested at the start of each chapter of this Teacher Pack, to enable teachers to detect prior knowledge before starting work on a topic.
- Probing questions are provided in each lesson plan to support informal assessment during the course of normal lessons.
- In order to facilitate regular monitoring of pupils' progress, at the end of each chapter of the Pupil Book, synoptic review questions are provided so that pupils and teachers have an opportunity to assess understanding of the topic covered.
- The 'Ready to progress' chart at the end of each chapter provides an opportunity for self-assessment, and for pupils to see what they need to do next in order to improve.
- Auto-marked homework tasks on the Collins Connect platform provide opportunities for formative assessment for each topic in the Pupil Books. Diagnostic feedback enables pupils to identify areas that they need to strengthen and directs them to resources that will help them. Results overviews help you to identify pupils' strengths and any areas where you might need to intervene.
- Tailor-made assessments are also provided for each tier of the *Maths Frameworking* scheme on the Collins Connect platform. A diagnostic assessment is available to evaluate pupils' starting points at the beginning of the first year of Key Stage 3 and to help you to decide how to set them. Thereafter, differentiated half-term tests are provided for each strand of the scheme, each one covering roughly five to six chapters of material. These tests evaluate pupils' progress against the stated objectives for each topic and again provide diagnostic feedback to identify pupils' strengths and areas for improvement.

Maths Frameworking and the new mathematics GCSE

In the new national curriculum, there is greater continuity between the Key Stage 3 Programme of study and the subject content of the new maths GCSE for first teaching in September 2015. The same six thematic strands are used in both Key Stages 3 and 4, and there are a number of examples where the wording used is the same in the subject content for both levels. This means that pupils who have a secure understanding of the material in the Key Stage 3 Programme of study will be very well prepared for the study of mathematics at Key Stage 4.

The assessment objectives for the new mathematics GCSE also build on and formalise the overarching aims of the Programme of study for Key Stages 1, 2 and 3, thereby forming a unified pathway for mathematics across primary and secondary. The Key Stage 3 Programme of study makes it clear, however, that there is no expectation that teaching of content from previous Key Stages should be repeated. Instead, pupils should be given the opportunity in Key Stage 4 to make choices and decisions about the mathematics they use, drawing on a well-developed toolkit to interpret and communicate mathematics for different audiences and purposes. It is this well-developed toolkit – a secure grounding in both content and process skills – that *Maths Frameworking 3rd edition* provides, thus creating the firm foundations that should enable every pupil to progress to GCSE and succeed at it.

Programme of study matching chart

This chart matches the subject content of the 2014 Programme of study for Key Stage 3 mathematics to specific lesson plans contained in this Teacher Pack.

Working mathematically Through the mathematics content, pupils should be taught to:	**Chapter title**	**Lesson number and title**
Number		
Understand and use place value for decimals, measures and integers of any size	9 Decimal numbers	9.2 Powers of 10 9.3 Rounding suitably 9.5 Solving problems
Order positive and negative integers, decimals and fractions; use the number line as a model for ordering of the real numbers; use the symbols =, ≠, <, >, ≤, ≥	Covered in Year 8	
Use the concepts and vocabulary of prime numbers, factors (or divisors), multiples, common factors, common multiples, highest common factor, lowest common multiple, prime factorisation, including using product notation and the unique factorisation property	Covered in Year 8	
Use the four operations, including formal written methods, applied to integers, decimals, proper and improper fractions, and mixed numbers, all both positive and negative	7 Fractions	7.1 Adding and subtracting fractions 7.2 Multiplying fractions 7.3 Dividing fractions
Use conventional notation for the priority of operations, including brackets, powers, roots and reciprocals	Covered in Year 7	
Recognise and use relationships between operations including inverse operations	1 Percentages	1.3 Calculating the original value 1.4 Using percentages
Use integer powers and associated real roots (square, cube and higher), recognise powers of 2, 3, 4, 5 and distinguish between exact representations of roots and their decimal approximations	Covered in Year 8	
Interpret and compare numbers in standard form $A \times 10^n$ $1 \le A < 10$, where n is a positive or negative integer or zero	Covered in Year 8	

Work interchangeably with terminating decimals and their corresponding fractions (such as 3.5 and $\frac{7}{2}$)	Covered in Year 8	
Define percentage as 'number of parts per hundred', interpret percentages and percentage changes as a fraction or a decimal, interpret these multiplicatively, express one quantity as a percentage of another, compare two quantities using percentages, and work with percentages greater than 100%	1 Percentages	1.1 Simple interest 1.2 Percentage increases and decreases 1.3 Calculating the original value
Interpret fractions and percentages as operators	1 Percentages	1.2 Percentage increases and decreases 1.3 Calculating the original value 1.4 Using percentages
Use standard units of mass, length, time, money and other measures, including with decimal quantities	12 Distance, speed and time	12.1 Distance 12.2 Speed 12.3 Time
Round numbers and measures to an appropriate degree of accuracy [for example, to a number of decimal places or significant figures]	9 Decimal numbers	9.1 Multiplication of decimals 9.3 Rounding suitably 9.4 Dividing decimals 9.5 Solving problems
Use approximation through rounding to estimate answers and calculate possible resulting errors expressed using inequality notation $a < x \leq b$	Covered in Year 8	
Use a calculator and other technologies to calculate results accurately and then interpret them appropriately	Covered in Year 7	
Appreciate the infinite nature of the sets of integers, real and rational numbers	Covered in Year 8	
Algebra		
Use and interpret algebraic notation, including: ab in place of $a \times b$ $3y$ in place of $y + y + y$ and $3 \times y$ a^2 in place of $a \times a$, a^3 in place of $a \times a \times a$; a^2b in place of $a \times a \times b$ $\frac{a}{b}$ in place of $a \div b$ coefficients written as fractions rather than as decimals brackets	2 Equations and formulae	2.4 Equations with fractions

Substitute numerical values into formulae and expressions, including scientific formulae	2 Equations and formulae	2.5 Formulae
Understand and use the concepts and vocabulary of expressions, equations, inequalities, terms and factors	Covered in Year 8	
Simplify and manipulate algebraic expressions to maintain equivalence by: collecting like terms multiplying a single term over a bracket taking out common factors expanding products of two or more binomials	2 Equations and formulae 8 Algebra	2.1 Multiplying out brackets 2.2 Factorising algebraic expressions 2.3 Equations with brackets 2.4 Equations with fractions 8.1 Expanding brackets 8.2 Factorising algebraic expressions 8.3 Expand and simplify
Understand and use standard mathematical formulae; rearrange formulae to change the subject	Covered in Year 8	
Model situations or procedures by translating them into algebraic expressions or formulae and by using graphs	Covered in Year 8	
Use algebraic methods to solve linear equations in one variable (including all forms that require rearrangement)	2 Equations and formulae	2.3 Equations with brackets 2.4 Equations with fractions
Work with coordinates in all four quadrants	Covered in Year 8	
Recognise, sketch and produce graphs of linear and quadratic functions of one variable with appropriate scaling, using equations in x and y and the Cartesian plane	Covered in Year 8	
Interpret mathematical relationships both algebraically and graphically	Covered in Year 8	
Reduce a given linear equation in two variables to the standard form $y = mx + c$; calculate and interpret gradients and intercepts of graphs of such linear equations numerically, graphically and algebraically	11 Solving equations graphically	11.1 Graphs from equations in the form $y = mx + c$ 11.2 Problems involving straight-line graphs

Use linear and quadratic graphs to estimate values of *y* for given values of *x* and vice versa and to find approximate solutions of simultaneous linear equations	11 Solving equations graphically	11.1 Graphs from equations in the form $y = mx + c$ 11.2 Problems involving straight-line graphs 11.3 Solving simple quadratic equations by drawing graphs 11.4 Problems involving quadratic graphs
Find approximate solutions to contextual problems from given graphs of a variety of functions, including piece-wise linear, exponential and reciprocal graphs	Covered in Year 8	
Generate terms of a sequence from either a term-to-term or a position-to-term rule	Covered in Year 8	
Recognise arithmetic sequences and find the *n*th term	Covered in Year 8	
Recognise geometric sequences and appreciate other sequences that arise	Covered in Year 8	
Ratio, proportion and rates of change		
Change freely between related standard units [for example time, length, area, volume/capacity, mass]	12 Distance, speed and time	12.1 Distance 12.2 Speed 12.3 Time
Use scale factors, scale diagrams and maps	6 Enlargements	6.1 Scale factors and enlargements 6.2 The centre of enlargement 6.3 Enlargements on grids
Express one quantity as a fraction of another, where the fraction is less than 1 and greater than 1	Covered in Year 7	
Use ratio notation, including reduction to simplest form	Covered in Year 8	
Divide a given quantity into two parts in a given part : part or part : whole ratio; express the division of a quantity into two parts as a ratio	Covered in Year 7	
Understand that a multiplicative relationship between two quantities can be expressed as a ratio or a fraction	Covered in Year 8	
Relate the language of ratios and the associated calculations to the arithmetic of fractions and to linear functions	Covered in Year 7	

Solve problems involving percentage change, including: percentage increase, decrease and original value problems and simple interest in financial mathematics	Covered in Year 7	
Solve problems involving direct and inverse proportion, including graphical and algebraic representations	Covered in Year 8	
Use compound units such as speed, unit pricing and density to solve problems	12 Distance, speed and time	12.1 Distance 12.2 Speed 12.3 Time
Geometry and measures		
Derive and apply formulae to calculate and solve problems involving: perimeter and area of triangles, parallelograms, trapezia, volume of cuboids (including cubes) and other prisms (including cylinders)	10 Surface area and volume of 3D shapes	10.1 Surface areas of cubes and cuboids 10.2 Volume formulae for cubes and cuboids 10.3 Volumes of triangular prisms
Calculate and solve problems involving: perimeters of 2D shapes (including circles), areas of circles and composite shapes	5 Circles 10 Surface area and volume of 3D shapes	5.1 The formula for the circumference of a circle 5.2 The formula for the area of a circle 5.3 Mixed problems 10.1 Surface areas of cubes and cuboids
Draw and measure line segments and angles in geometric figures, including interpreting scale drawings	Covered in Year 7	
Derive and use the standard ruler and compass constructions (perpendicular bisector of a line segment, constructing a perpendicular to a given line from/at a given point, bisecting a given angle); recognise and use the perpendicular distance from a point to a line as the shortest distance to the line	Covered in Year 8	
Describe, sketch and draw using conventional terms and notations: points, lines, parallel lines, perpendicular lines, right angles, regular polygons, and other polygons that are reflectively and rotationally symmetric	Covered in Year 8	

Use the standard conventions for labelling the sides and angles of triangle ABC, and know and use the criteria for congruence of triangles	6 Enlargements	6.1 Scale factors and enlargements 6.2 The centre of enlargement 6.3 Enlargements on grids
Derive and illustrate properties of triangles, quadrilaterals, circles, and other plane figures [for example, equal lengths and angles] using appropriate language and technologies	Covered in Year 8	
Identify properties of, and describe the results of, translations, rotations and reflections applied to given figures	3 Polygons	3.1 Polygons
Identify and construct congruent triangles, and construct similar shapes by enlargement, with and without coordinate grids	6 Enlargements	6.1 Scale factors and enlargements 6.2 The centre of enlargement 6.3 Enlargements on grids
Apply the properties of angles at a point, angles at a point on a straight line, vertically opposite angles	Covered in Year 8	
Understand and use the relationship between parallel lines and alternate and corresponding angles	Covered in Year 8	
Derive and use the sum of angles in a triangle and use it to deduce the angle sum in any polygon, and to derive properties of regular polygons	3 Polygons	3.2 Angles in polygons 3.3 Interior angles of regular polygons
Apply angle facts, triangle congruence, similarity and properties of quadrilaterals to derive results about angles and sides, including Pythagoras' Theorem, and use known results to obtain simple proofs	Covered in Year 8	
Use Pythagoras' Theorem and trigonometric ratios in similar triangles to solve problems involving right-angled triangles	13 Similar triangles	13.1 Similar triangles 13.2 A summary of similar triangles 13.3 Using triangles to solve problems
Use the properties of faces, surfaces, edges and vertices of cubes, cuboids, prisms, cylinders, pyramids, cones and spheres to solve problems in 3D	10 Surface and volume of 3D shapes	10.1 Surface areas of cubes and cuboids 10.2 Volume formulae for cubes and cuboids 10.3 Volumes of triangular prisms
Interpret mathematical relationships both algebraically and geometrically	Covered in Year 8	

Probability		
Record, describe and analyse the frequency of outcomes of simple probability experiments involving randomness, fairness, equally and unequally likely outcomes, using appropriate language and the 0–1 probability scale	Covered in Year 8	
Understand that the probabilities of all possible outcomes sum to 1	Covered in Year 8	
Enumerate sets and unions/intersections of sets systematically, using tables, grids and Venn diagrams	Covered in Year 8	
Generate theoretical sample spaces for single and combined events with equally likely, mutually exclusive outcomes and use these to calculate theoretical probabilities	Covered in Year 8	
Statistics		
Describe, interpret and compare observed distributions of a single variable through: appropriate graphical representation involving discrete, continuous and grouped data; and appropriate measures of central tendency (mean, mode, median) and spread (range, consideration of outliers)	Covered in Year 8	
Construct and interpret appropriate tables, charts, and diagrams, including frequency tables, bar charts, pie charts, and pictograms for categorical data, and vertical line (or bar) charts for ungrouped and grouped numerical data	4 Using data	4.3 Two-way tables 4.4 Comparing two or more sets of data 4.5 Statistical investigations
Describe simple mathematical relationships between two variables (bivariate data) in observational and experimental contexts and illustrate using scatter graphs	4 Using data	4.1 Scatter graphs and correlation 4.2 Interpreting graphs and diagrams 4.4 Comparing two or more sets of data

1 Percentages

Learning objectives

- How to calculate simple interest
- How to use a multiplier to calculate percentage increases and decreases
- How to calculate the original value after a percentage change

Prior knowledge

- How to work out a percentage of a given number, with or without a calculator
- How to write one number as a percentage of another number

Context

- Percentage increase and decrease is probably one of the most common uses of mathematics in real life. Everyone meets it in some form or other, even if only in terms of financial capability. You could use the following link to find real-life applications of percentage: **http://www.pfeg.org/**

Discussion points

- Which sets of equivalent fractions, decimals and percentages do you know?
- From one set that you know, for example: $\frac{1}{2} = 0.5 = 50\%$, which others can you work out?
- How would you go about finding the percentage equivalents of any fraction?

Associated Collins ICT resources

- Chapter 1 interactive activities on Collins Connect online platform
- *Calculating reverse percentages* video on Collins Connect online platform
- *Cars and phones* Wonder of Maths on Collins Connect online platform

Curriculum references

Develop fluency

- Select and use appropriate calculation strategies to solve increasingly complex problems

Solve problems

- Develop their use of formal mathematical knowledge to interpret and solve problems, including in financial mathematics

Number

- Define percentage as 'number of parts per hundred', interpret percentages and percentage changes as a fraction or a decimal, interpret these multiplicatively, express one quantity as a percentage of another, compare two quantities using percentages, and work with percentages greater than 100%
- Interpret fractions and percentages as operators
- Recognise and use relationships between operations including inverse operations

Fast-track for classes following a 2-year scheme of work

- Although pupils have met percentages before there are some important and quite challenging concepts in this chapter for pupils working at this level. The ideas of percentages as a multiplier and the use of multiplicative reasoning are very important to pupils' confidence and fluency when working with percentages. So, while you may be able to leave out some of the earlier questions in each exercise, be careful not to leave out too much or move on too fast.

Lesson 1.1 Simple interest

Learning objectives

- To understand what simple interest is
- To solve problems involving simple interest

Resources and homework

- Pupil Book 3.1, pages 7–9
- Intervention Workbook 2, pages 40–42
- Intervention Workbook 3, pages 16–17
- Homework Book 3, section 1.1
- Online homework 1.1, questions 1–10

Links to other subjects

- **History** – to calculate the impact of interest over time to compare incomes or standards of living

Key words

- lender
- simple interest

Problem solving and reasoning help

- **MR** questions 10 and 11 in Exercise 1A of the Pupil Book require pupils to apply their understanding to more complex financial situations (you could use question 10 to start pupils thinking about compound interest). The challenge at the end of the exercise requires pupils to think about how they can use formulae to help them generalise questions about percentage change. This is a powerful idea that you will need to develop carefully with the class.

Common misconceptions and remediation

- Pupils often struggle when they start using percentages that are greater than 100. Using a real-life example will help pupils to overcome this. Start with percentages that pupils can work with comfortably, for example, an explanation based on a shop selling a pair of jeans that cost £30 to make. A 50% profit would mean that the shop sold the jeans for £30 plus £15, which is £45; 100% profit would mean selling the jeans for £30 plus £30, which is £60; 150% profit would mean selling the jeans for £30 plus £45, which is £75.

Probing questions

- Talk me through how you would increase or decrease £22 by, for example, 25%.
- Can you work it out in a different way?

Part 1

- Working in pairs, ask pupils to discuss what they know about taking out a loan, and to suggest some examples of when they might take out a loan.
- Then ask pupils to discuss what they know about how they might pay back the loan.
- Take feedback from pupils, but do not comment on the feedback yet.

Part 2

- Say that at some point most people will need to take out a loan, for example, to buy a house. Interest is usually paid to the *lender* – the company or person that has provided the loan.
- This may be a good point to comment on any suggestions raised here.
- One type of interest is called *simple interest*. This is calculated as a percentage. As long as the loan exists, the recipient will pay the lender a percentage of the loan at regular intervals.
- Work through examples 1 and 2 on page 7 of the Pupil Book as a class. Check pupils' understanding before they do Exercise 1A.
- **Pupils can now do Exercise 1A from Pupil Book 3.1.**
- You could ask **more able** pupils to use the internet to find out about compound interest.

Part 3

- Working in pairs, ask pupils to plan a short presentation to talk through how you would increase and/or decrease £12 by, for example, 15%. Can you do it in a different way? How would you find the multiplier for different percentage increases/decreases?
- Chose pupils to present their explanation to the class. The class should provide formative feedback on an agreed set of criteria such as: clarity and accuracy of explanation, use of examples and different methods. This will prepare pupils for Lesson 1.2 and Lesson 1.3.

Answers

Exercise 1A

1	**a** £1.50	**b** £3.50	**c** £8	**d** £21
	e £12	**f** £22	**g** £24	**h** £6
2	**a** £2.25	**b** £1.75	**c** £1.05	**d** £2.15
	e £20.80	**f** £43.20	**g** £4.80	**h** £42.60
3	**a** £30	**b** £52.50		
4	**a** £60	**b** £130		
5	**a** £135	**b** £4770		
6	**a** £51.10	**b** £883.30		
7	£5544			
8	**a** £108	**b** £408	**c** £34	
9	**a** £810	**b** £1560	**c** £130	

d pupils' own answers; for example, a loan shark as the interest rates are very high

10 Gabriel pays more (£12 × 12 = £144) than Joshua (£120).

11 yes, 0.5% × 12 = 6%

Challenge: Using a formula

A £150

B **a** £56 **b** £360 **c** £2800

Lesson 1.2 Percentage increases and decreases

Learning objectives
- To calculate the result of a percentage increase or decrease
- To choose the most appropriate method to calculate a percentage change

Resources and homework
- Pupil Book 3.1, pages 10–13
- Intervention Workbook 2, pages 40–42
- Intervention Workbook 3, pages 16–17
- Homework Book 3, section 1.2
- Online homework 1.2, questions 1–10

Links to other subjects
- **Food technology** – to calculate increases in a certain food type in the diets of different people

Key words
- decrease
- multiplier
- increase

Problem solving and reasoning help
- This lesson reinforces the concept of using percentage as an operator. This is an important step to ensure confidence and fluency in pupils, so make sure that you spend time on this lesson. It often helps to make links to fractions as operators. In the challenge at the end of Exercise 1B in the Pupil Book, pupils will need to apply their understanding of percentage change to the slightly less familiar context of population change. Use the opportunity to tackle any outstanding misconceptions.

Common misconceptions and remediation
- Pupils are often confused when they come across percentages that are greater than 100. Using real-life examples could help, starting with percentages that pupils are able to work with comfortably. Pupils need a good understanding of 100% as a whole before tackling percentage increases and decreases successfully.

Probing questions
- How would you find the multiplier for different percentage increases and decreases?
- How would you find a multiplier to increase, then decrease by a given percentage?
- Given a multiplier, how could you tell if this would result in an increase or a decrease?

Part 1
- Write a variety of percentages on the board such as: 5%, 10%, 20%, 25%. Then write some quantities on the board, for example: £32, 58 kg, 200 km, £150.
- Apply each percentage value to each quantity and calculate a percentage increase or decrease, as appropriate. Pupils should be able to do most of these without a calculator, but let them decide which percentages they can do easily without a calculator.
- In pairs, encourage pupils to challenge each other to calculate easier percentages faster than when her or his partner is using a calculator.

Part 2
- Tell pupils that a percentage change may be:
  - an *increase* if the new value is larger than the original value
  - a *decrease* if the new value is smaller than the original value.
- Say that there are several methods to use to calculate the result of a percentage change. Ask pupils if they used other methods of calculation in Part 1. If so, ask these pupils to explain the differences between methods.

- Tell pupils that the *multiplier* method is often the most efficient, and the focus of this lesson.
- Say that the multiplier method involves multiplying the original value by an appropriate number to calculate the result of the percentage change.
- Work through Example 3 on page 10 of the Pupil Book as a class; then to work through examples 4 to 6 in pairs. Check pupils' understanding before they do Exercise 1B.
- **Pupils can now do Exercise 1B from Pupil Book 3.1**.

Part 3

- Ask pupils to work in pairs on the following problem: 'The answer to a percentage increase question is £10. Make up an easy question and one that is difficult.'
- Pairs should share their questions with another pair.
- As a class, summarise by discussing what makes the questions easy or difficult.

Answers

Exercise 1B

1 **a** 1.04 **b** 1.06 **c** 1.09 **d** 1.1 **e** 1.12
f 1.15 **g** 1.2 **h** 1.35 **i** 1.17

2 **a** 0.98 **b** 0.97 **c** 0.95 **d** 0.92 **e** 0.9
f 0.85 **g** 0.8 **h** 0.7 **i** 0.75

3 **a** £13.20 **b** £21 **c** £31.20 **d** £108 **e** £224

4 **a** £16.20 **b** £29.10 **c** £47 **d** £135 **e** £320

5 **a** 96 cm **b** 1 m 15 cm (115.2 cm)

6 **a** 49.5 kg **b** 54.45 kg

7 **a** £530 **b** £561.80

8 **a** 2.4 mg **b** 2.88 mg

9 **a** 13 500 **b** 14 580

10 **a** 300 megalitres **b** 180 megalitres

11 **a** £10 200 **b** £8670

12 £124 848

13 × 1.1 × 0.9 = 0.99 so a 1% decrease overall

Challenge: Population change

A 5760

B 5760 × 1.2 = 6912

C 8294

D 2017 (11 944)

Lesson 1.3 Calculating the original value

Learning objective

- Given the result of a percentage change, to calculate the original value

Resources and homework

- Pupil Book 3.1, pages 14–16
- Intervention Workbook 2, pages 40–42
- Intervention Workbook 3, pages 16–17
- Homework Book 3, section 1.3
- Online homework 1.3, questions 1–10

Links to other subjects

- **Food technology** – to calculate comparative dietary components
- **Geography** – to calculate populations at a given period, given the current size of the population and a rate of change

Key words

- original value

Problem solving and reasoning help

- This lesson continues to develop the concept of using percentage as an operator by looking at the inverse to calculate the percentage change or to calculate an initial value. Pupils will need to be fluent with this concept so that they are confident in applying their understanding of percentages to real-life problems. Encourage discussion to challenge any misconceptions.

Common misconceptions and remediation

- Pupils are often confused by percentages that are greater than 100. Pupils also make the mistake of pairing an increase with an equivalent decrease. For example, they expect a 50% increase that is followed by a 50% decrease to return them to the starting value. This misconception results from using additive instead of multiplicative reasoning. This lesson tackles this misconception. In Part 3 pupils should draw on the work they did on inverse relationships during the lesson to check and consolidate their understanding.

Probing questions

- After the addition of 8% interest, Sarah has savings of £850. What was the original amount?
- After one year a car depreciates by 5% and is valued at £8500. What was the value of the car at the beginning of the year?
- After being given a 2% increase, a shop assistant earns £7.50 per hour. What was his hourly rate before the increase?
- At a 25% discount sale, a boy pays £30 for a pair of jeans. What was the original price?
- Explain how to find a multiplier to calculate an original value after a proportional increase or decrease.
- True or false: 'The inverse of an increase by a percentage is not a decrease by the same percentage.' Justify your answer.

Part 1

- Allow five minutes for individuals to do the following question; then five to discuss it in pairs.
  - At a sale, prices were reduced by 33%. After the sale, prices were increased by 50%.
  - What was the overall effect on the prices? Explain how you know.
- Say that you will revisit this at the end of the lesson to see who has changed their mind.

Part 2

- Use examples 7 to 9 on pages 14 and 15 of the Pupil Book to demonstrate how to use multiplicative reasoning and inverse operations to calculate the percentage change between two values. This concept will challenge pupils working at this level, so make sure that you take time over it in order to give pupils the opportunity to ask questions.
- If necessary, support pupils with more worked examples as part of guided group work.
- **Pupils can now do Exercise 1C from Pupil Book 3.1.**

Part 3

- Revisit the question in Part 1. Encourage pupils to use what they have done on percentages as a multiplier and the use of inverse relationships to explain their response to this question.
- Extension: What percentage increase would take the sale price back to its starting point?

Answers

Exercise 1C

1	**a** 1.05	**b** 1.07	**c** 1.11	**d** 1.13	**e** 1.18
	f 1.25	**g** 1.4	**h** 1.55	**i** 1.34	
2	**a** 0.96	**b** 0.97	**c** 0.95	**d** 0.92	**e** 0.9
	f 0.85	**g** 0.8	**h** 0.7	**i** 0.55	
3	**a** £30	**b** £16	**c** £36	**d** £44	**e** £120
4	**a** £60	**b** £70	**c** £80	**d** £50.50	**e** £10.80

5 **a** 100% + 5% = 105% = 105 ÷ 100 = 1.05
b original height × 1.05 = new height, so original height = new height ÷ 1.05
c 120 cm

6 **a** 100% + 10% = 110% = 110 ÷ 100 = 1.1
b mass at 1 month old × 1.1 = mass at 2 months old, so mass at 1 month old = mass at 2 months old ÷ 1.1
c 5.5 kg **d** 5 kg

Challenge: Trees

A 5.0625 m
B 1 m
C after 6 years

Lesson 1.4 Using percentages

Learning objectives

- To revise the links within fractions, decimals and percentages
- To choose the correct calculation to work out a percentage

Resources and homework

- Pupil Book 3.1, pages 17–21
- Intervention Workbook 2, pages 40–42
- Intervention Workbook 3, pages 16–17
- Homework Book 3, section 1.4
- Online homework 1.4, questions 1–10

Links to other subjects

- **Food technology** – to calculate dietary requirements for different people
- **Geography** – to calculate comparative populations in different areas or over time

Key words

- No new key words for this topic

Problem solving and reasoning help

- This lesson develops and consolidates pupils' understanding of percentages by enabling them to make choices and decisions about the methods they use in a range of contexts.

Common misconceptions and remediation

- Pupils often learn mathematics rules without really understanding them. Thus, particularly at this level, questions come up over time and pupils may not have had the chance to identify the type of question and make independent decisions about which method to use to solve it. Provide opportunities for pupils to check their understanding by choosing and deciding which approach they should use, in a range of increasingly complex and unfamiliar situations.

Probing questions

- The answer to a percentage increase question is £12. Make up an easy and a difficult question.

Part 1

- This activity recaps earlier learning. Write four quantities on the left-hand side of the board, for example: 45, 60, 56, 12. On the right-hand side write, for example: 120, 300, 200, 160.
- Match the quantities; one from the left-hand side and one from the right-hand side.
- Calculate, with or without a calculator, the percentage that the left value is of the right value.
- You may want to provide a worked example for pupils working at this level.
- Some pairs are obvious (12 and 120); some pairs are clearly non-calculator (45 out of 300).
- Some pairs have whole-number answers, but are not obvious such as 56 out of 160 (35%). Discuss the appropriate methods to use for the answers that are not obvious.

Part 2

- Tell pupils that now that they know how to answer a range of percentage questions, they need to be able to decide on the type of question and thus the method to use to solve it. Make it clear that pupils will need to draw on all their learning from this chapter, what they already know about percentages, as well as make links to fractions, decimals and ratio.
- Write the questions from Example 10 on page 18 of the Pupil Book on the board:
  - In a group, the ratio of boys to girls is 7 : 3.
  - What percentage of the group is made up of boys?

- Working in pairs, ask pupils to describe in their own words what type of questions these are and what mathematics they will have to use. Encourage them to identify key words that will help them to decide what to do. Pupils do not need to work out the answer yet.
- Take feedback (pupils share key words). As a class, look at the solution in the Pupil Book.
- This is challenging, but very important in terms of pupils being able to apply their learning independently. Ensure that you allow time for pupils to ask questions.
- Go through a similar process with pupils for Example 11 and 12.
- Give plenty of time for questions and encourage pupils to identify what is the same or different about the questions and try to categorise the different types of questions.
- If necessary, support pupils with more worked examples as part of guided group work. You may need to provide some simple examples to help pupils review what they already know.
- **Pupils can now do Exercise 1D from Pupil Book 3.1.**

Part 3

- Revisit what you did at the end of Lesson 1.2, when pupils decided what made a percentage increase question easy or difficult. Tell pupils that now they will extend this, using all they know about percentages as well as fractions, decimals and ratio to design their own percentage questions. Ask pupils to make up an easy question and one that is difficult.
- Share some examples with the class and discuss what makes a question easy or difficult.
- Use grouping to support or challenge pupils.

Answers

Exercise 1D

1 **a** 0.75 **b** 0.8 **c** 0.8 **d** 0.875
e 0.35 **f** 0.36 **g** 0.37 **h** 0.16

2 **a** 75% **b** 85% **c** 38% **d** 17%
e 70% **f** 90% **g** 15.5% **h** 75.5%

3 **a** 80% **b** 75% **c** 62.5% **d** 27%
e 62% **f** 35% **g** 87.5%

4 $\frac{1}{2} = 0.5 = 50\%$
$\frac{1}{4} = 0.25 = 25\%$
$\frac{7}{8} = 0.875 = 87.5\%$
$\frac{3}{5} = 0.6 = 60\%$
$\frac{7}{10} = 0.7 = 70\%$
$\frac{13}{20} = 0.65 = 65\%$

5 64%, 90%, 60%, 60%, 70%

6 **a** 70%, 95%, 70%, 67.5%, 80%
b onions, leeks and cauliflower, as none of these is 80% or more

7 **a i** $\frac{20}{500} = \frac{1}{25}$ **ii** 0.04 **iii** 4%
b 25.6%
c The total number of pupils is 100, so the percentage is 3%.

8 Sophia is correct, as $\frac{17}{20} = \frac{85}{100} = 85\%$.

9 40%

10 **a** 200 **b i** $\frac{3}{10}$ **ii** 30%
c 16% **d** 54%

Challenge: Different representations

A Check pupils' answers.

B for example, 54% of the year are boys or the fraction of pupils in the year who are girls is $\frac{46}{100}$

C for example, 48% of the school are girls or the fraction of pupils in the school who are boys is $\frac{468}{900}$

Review questions

(Pupil Book pages 22–23)

- The review questions will help to determine pupils' abilities with regard to the material within Chapter 1.
- The answers are on the next page of this Teacher Pack.

Challenge – The Royal Albert Hall

(Pupil Book pages 24–25)

- This challenge gives pupils the opportunity to extend their learning to a real-life context.
- All the information pupils will need is provided in the Pupil Book, but it is quite complex. Pupils working at this level may find it difficult to access the information they need. This is representative of how they are likely to find information presented in real life. Pupils will need to read the questions very carefully to decide what information they need and what mathematical skills to use in each case.
- You may want to discuss the information provided as a class or in guided groups.
- As a warm-up to this activity, you could ask pupils to explore the websites of some local venues.
- You could ask pupils to answer questions related to the pricing on these sites. You could also use 'What if?' questions that involve percentage changes. (You may want to check pupils' understanding of the multiplier method for percentage increase and decrease, which they met in Lesson 1.2.)
- Ask pupils to answer questions 1 to 6 of this challenge activity on page 25 of the Pupil Book.
- As an extension, you could ask pupils to design some questions of their own.

Answers to Review questions

1	**a** £0.60	**b** £2	**c** £7.50	**d** £3.80	
	e £7.20	**f** £36	**g** £29.40	**h** £41	
2	**a** £48	**b** £70.67			
3	**a** 1.03	**b** 1.05	**c** 1.08	**d** 1.11	**e** 1.14
	f 1.18	**g** 1.22	**h** 1.45	**i** 1.33	
4	**a** 0.96	**b** 0.98	**c** 0.93	**d** 0.91	**e** 0.88
	f 0.82	**g** 0.75	**h** 0.65	**i** 0.67	
5	**a** £8.32	**b** £31.20	**c** £21	**d** £53	**e** £345
6	**a** £19	**b** £38.40	**c** £55.20	**d** £225	**e** £264
7	**a** 30 kg	**b** 36 kg			
8	**a** 60%	**b** 25%	**c** 37.5%	**d** 39%	**e** 84%
	f 45%	**g** 62.5%			
9	**a** £11 200	**b** £8960			
10	**a** £576	**b** £1176	**c** £98		
	d for example: a loan shark as the interest paid is very high				
11	**a** £30	**b** £14	**c** £32	**d** £39	**e** £110
12	**a** 100% + 4% = 104% = 104 ÷ 100 = 1.04				
	b original height × 1.04 = new height so original height = new height ÷ 1.04				
	c 125 cm				
13	**a** 200	**b i** 1/5	**ii** 20%		
	c 14%	**d** 66%			
14	**a** £50	**b** £30	**c** £40	**d** £60	**e** £20.30

Answers to Challenge – The Royal Albert Hall

1 **a** 1971 **b** 2021

2 4400

3 **a** 9 hours **b** 16 443 hours

4 £18 600 000

5 £27.2 million

6 Rounding the values gives 2 000 000 × 100 = 200 000 000, which is less than a billion, so Helen is right.

2 Equations and formulae

Learning objectives

- How to expand brackets and factorise algebraic expressions
- How to solve equations
- How to use formulae

Prior knowledge

- How to collect like terms in an expression
- How to use one or two operations to solve equations
- How to substitute values into a formula
- What a highest common factor (HCF) is

Context

- This chapter builds on previously learned algebraic techniques and moves on to more advanced methods of algebraic manipulation. These include: simplifying expressions and expanding brackets, factorising algebraic expressions, solving linear equations involving fractions and using formulae.

Discussion points

- What steps do you follow when expanding a bracket? What happens if the bracket has a negative coefficient?
- What is a variable and why do we use them?
- What strategies can be used to solve for unknowns in algebraic equations?
- Why do we need to be able to rearrange formulae?

Associated Collins ICT resources

- Chapter 2 interactive activities on Collins Connect online platform

Curriculum references

Develop fluency

- Substitute values in expressions, rearrange and simplify expressions, and solve equations

Solve problems

- Develop their mathematical knowledge, in part through solving problems and evaluating the outcomes, including multi-step problems

Algebra

- Use and interpret algebraic notation, including:
  - coefficients written as fractions rather than as decimals
  - brackets
- Substitute numerical values into formulae and expressions, including scientific formulae

- Simplify and manipulate algebraic expressions to maintain equivalence by:
    - collecting like terms
    - multiplying a single term over a bracket
    - taking out common factors
    - expanding products of two or more binomials
- Use algebraic methods to solve linear equations in one variable (including all forms that require rearrangement)

Fast-track for classes following a 2-year scheme of work

- Much of this chapter will be unfamiliar to pupils. However, some pupils may be familiar with expanding brackets. Check that all pupils can expand brackets fluently before moving on to the rest of the chapter. If pupils grasp the concepts quickly they can move on to the more challenging questions that are towards the end of each exercise in the Pupil Book.

Lesson 2.1 Multiplying out brackets

Learning objective

- To multiply out brackets

Resources and homework

- Pupil Book 3.1, pages 27–29
- Homework Book 3, section 2.1
- Online homework 2.1, questions 1–10

Links to other subjects

- **IT** – to know theoretical computer science and to use formulae to break codes

Key word

- expand

Problem solving and reasoning help

- The challenge at the end of Exercise 2A in the Pupil Book enables pupils to demonstrate their learning. Encourage **less able** pupils to follow the methods shown in the Pupil Book in order to work out the answers to each question. You could ask **more able** pupils to design their own code-breaking questions.

Common misconceptions and remediation

- A common problem when expanding a bracket is for pupils to multiply the first term by the number outside the bracket and just write down the second term. Also, pupils sometimes add instead of multiplying. Use Part 1 of this lesson to make sure that pupils can expand brackets correctly before they tackle Exercise 2A.

Probing questions

- Show pupils examples of errors when multiplying out brackets, and ask them to point out and explain the errors: $3(x + 2) = 3x + 2$; $2(x - 3) = 2x - 5$; $-2(6 - x) = -12 - 2x$

Part 1

- Ask how many pounds make one kilogram. Accept 2 as the answer to start with (if a pound is 500 g). Work towards the correct answer, which is 2.2 pounds (450 g or 454 g per pound).
- Now ask what the equivalent of 6 kg is in pounds. Ask pupils to explain how they worked this out, probably: $6 \times 2 = 12$ added to $6 \times 0.2 = 1.2$, giving 13.2.
- Using a target board such as the one shown, work your way around the class asking pupils for pound equivalents of kilogram weights.

8	5	15	4	7
3	12	9	11	20
10	13	100	2	3

Part 2

- Draw a rectangle on the board with the dimensions 4 cm and 5 cm. Ask the class to give you the area of the rectangle. When someone gives you the correct answer (20 cm^2), ask the pupil how she or he calculated it. (4 cm × 5 cm)
- Draw a second rectangle on the board with the dimensions $4x$ and $3x$. Ask for the area of the rectangle. This should create some discussion, leading to the correct answer of $12x$.
- Now draw a rectangle with the dimensions $3x + 2$ and 5 on the board. Ask pupils for the area of the rectangle. Some pupils will see the correct answer ($15x + 10$) more easily than others.
- Show that the last example is the same as $5(3x + 2)$. Show that we expand the brackets by multiplying each term to arrive at $15x + 10$.
- Now write the expression $3(4x + 2)$ on the board and ask the class what this might mean. You are looking for the response: 'the area of a rectangle with dimensions 3 and $4x + 2$' as well as the expansion of the brackets to give $12x + 6$.

- Repeat the process for $x(3x + 2)$ to give $3x^2 + 2x$.
- **Pupils can now do Exercise 2A from Pupil Book 3.1.**

Part 3

- Ask pupils to expand the brackets of these expressions:

  $3(x + 4)$ $2(x - 5)$ $3(2x + 4)$ $\frac{1}{3}(6x - 3)$

Answers

Exercise 2A

1 **a** $7x$ **b** $7a$ **c** $8t$ **d** $8y$
e $6m$ **f** $3k$ **g** $4n$ **h** $-4p$

2 **a** $10m$ **b** $7y$ **c** $9t$ **d** $11p$
e $13n$ **f** $9p$ **g** $6t$ **h** $7e$
i $6k$ **j** $5h$ **k** $5m$ **l** $6t$

3 **a** $P = 8T$ **b** $P = 8N$ **c** $P = 14m$
d $P = 13k$ **e** $P = 18w$ **f** $P = 13n$

4 **a** $5b + 3$ **b** $7x + 6$ **c** $6q + 3$ **d** $7k + 7$
e $2x + 5$ **f** $6k + 3$ **g** $2p + 1$ **h** $3d + 2$
i $2m - 2$ **j** $2t - 3$ **k** $3w - 7$ **l** $4g - 5$
m $7t + k$ **n** $9x + 3y$ **o** $7k + 2g$ **p** $5h + 4w$
q $3t + 3p$ **r** $2n + 3t$ **s** $p + q$ **t** $2n + p$

5 **a** $8x$ **b** $12a$ **c** $10t$ **d** $6y$ **e** $24k$
f $15t$ **g** $12x$ **h** $12m$ **i** $12t$ **j** $35y$

6 **a** $3t + 12$ **b** $3x + 15$ **c** $2m - 6$ **d** $4k - 8$
e $6 + 2x$ **f** $12 - 3k$ **g** $24 - 4y$ **h** $15 - 5x$

7 **a** $A = 5x + 15$ **b** $A = 3t + 6$ **c** $A = 2m - 2$ **d** $A = 20 + 4k$

8 **a** $2m + 6$ **b** $3k - 12$ **c** $3a + 6$ **d** $15 - 5p$
e $6x + 8$ **f** $10x + 15$ **g** $8t - 4$ **h** $20m + 35$
i $6x + 3$ **j** $12k - 8$ **k** $10b + 6$ **l** $14 - 28m$
m $24 + 8p$ **n** $20 - 5t$ **o** $12 - 18g$ **p** $16 + 24t$
q $18k - 54$ **r** $10m + 15$ **s** $9t - 6$ **t** $6 - 8y$

9 **a** $A = 3x + 6$ **b** $A = 4x + 10$ **c** $A = 15m + 20$
d $A = 35k + 7p$ **e** $A = 12t + 8$ **f** $A = 6x + 15$

10 $3(4x - 2) = 12x - 6$ and $6(2x - 1) = 12x - 6$

11 He has only multiplied the 5 by the $2x$. He should also multiply it by the 3.

Challenge: Code breaker

A–G pupils' own work

H APRIL FOOL

Lesson 2.2 Factorising algebraic expressions

Learning objective

- To factorise expressions

Resources and homework

- Pupil Book 3.2, pages 30–32
- Homework Book 3, section 2.2
- Online homework 2.2, questions 1–10

Links to other subjects

- **Business studies** – to be able to do financial calculations

Key word

- factorise

Problem solving and reasoning help

- Remind pupils to factorise an expression completely in order to get the correct answer.
- Explain that factorising $4x - 20$ to give $2(2x - 10)$ has not been completely factorised, as the HCF of 4 and 20 is 4, not 2. Completely factorising the expression would give: $4(x - 5)$.

Common misconceptions and remediation

- Pupils often take out only the factor from the first part of an expression, for example: $4x - 10 = 2(2x - 10)$. Remind pupils to take out the factor from *all* elements of the expression.

Probing questions

- What are the steps that you need to take when factorising a linear expression?
- Show pupils examples of errors that can be made when factorising expressions. Ask pupils to explain what the errors are in each of these expressions:

  $3x + 9 = 3(x + 9)$ $\qquad$ $2x - 12 = 2(2x - 6)$ $\qquad$ $-12 - 2x = 2(6 - x)$.

Part 1

- A formula for the approximate conversion of temperatures from degrees Fahrenheit to degrees Celsius is $C = \frac{1}{2}(F - 32)$, where C is the temperature in degrees Celsius and F is the temperature in degrees Fahrenheit.
- Ask the class to use this formula to estimate the equivalent of 100 °Fahrenheit in °Celsius (34). Discuss with the class the strategy they used to do this calculation mentally.
- Work through more examples such as: 66 – 30 $\quad$ 36 – 2 $\quad$ 34 ÷ 2 $\quad$ 17

  90 – 30 $\quad$ 60 – 2 $\quad$ 58 ÷ 2 $\quad$ 29
- Using a target board as shown, work your way around the class asking pupils to convert temperatures in °Fahrenheit to their approximate equivalents in °Celsius.

34	109	38	40	73
55	42	76	89	50
61	32	57	71	88
99	93	103	67	72

Part 2

- On the board, draw a rectangle with the area of 7 cm^2 written inside it. Ask the class what dimensions the rectangle could have. The simplest answer is 1 cm by 7 cm. Explain that coming up with these numbers involves finding a pair of factors.
- Now draw a rectangle with the area of 12 cm^2 written inside it. What dimensions could the rectangle have? (More than one choice; any factor pair will do: 2 cm by 6 cm, 3 cm by 4 cm)
- Draw a rectangle with the expression $3x + 3$ written inside it. Tell pupils that this is the area of the rectangle and ask what dimensions the rectangle could have. Lead the discussion so that the class reaches the correct answer of 3 and $x + 1$. Say that to reach this answer, we

must find a pair of factors, which, when multiplied together give $3x + 3$. Explain that this is called *factorising*, and is the opposite process of expansion (covered in Lesson 2.1).

- Write the expression: $6x + 9$. Say that this is the area of a rectangle. What dimensions could the rectangle have? Pupils need to factorise the expression to create a bracket with a term outside the two factors. Here, factorisation gives $3(2x + 3)$. This expands to give $6x + 9$.
- Write $x^2 + 5x$ on the board. Ask what dimensions a rectangle of this area could have. Help pupils to see that this factorisation will be $x(x + 5)$; expansion will give the original: $x^2 + 5x$.
- **Pupils can now do Exercise 2B from Pupil Book 3.1.**

Part 3

- Ask pupils what is meant by factorisation. You want responses that show understanding of breaking down an expression into two terms that will multiply together to give the original.
- Review the two stages pupils have gone through in this lesson with examples such as:

  $6 + 9x = 3(2 + 3x)$ $\qquad$ $5x^2 - 3x = x(5x - 3)$

Answers

Exercise 2B

1 **a** 1, 2, 3, 4, 6, 12 **b** 1, 3, 5, 15 **c** 1, 2, 3, 6, 9, 18 **d** 1, 2, 4, 5, 10, 20
e 1, 2, 3, 4, 6, 8, 12, 24 **f** 1, 2, 5, 10 **g** 1, 2, 4, 8 **h** 1, 2, 3, 5, 6, 10, 15, 30

2 **a** 2, 3, 4, 12 **b** 5, 10, 20, 35 **c** 2, 10, 12, 20, 4, 8, 18, 24
d 5, 35 **e** 3, 5, 11, 7, 21, 35 **f** 2, 5, 10, 20, 4, 8

3 **a** 2 **b** 3 **c** 6 **d** 6

4 **a** $3(x + 2)$ **b** $2(t + 3)$ **c** $4(n + 2)$ **d** $2(q + 4)$
e $3(x - 3)$ **f** $4(p - 1)$ **g** $5(y - 2)$ **h** $3(t - 4)$
i $2(4 + x)$ **j** $4(3 + k)$ **k** $6(2 - t)$ **l** $3(5 - k)$

5 **a** $3(t + 3)$ **b** $2(m + 2)$ **c** $5(p + 1)$ **d** $4(m + 3)$
e $6(k - 3)$ **f** $3(n - 2)$ **g** $2(x - 4)$ **h** $3(q - 5)$
i $5(2 + x)$ **j** $4(4 + h)$ **k** $3(4 - t)$ **l** $6(3 - k)$

6 **a** 4 **b** $2t + 1$ **c** 2 **d** $3 - 2y$

7 **a** $4(x + 2)$ **b** $6(t + 2)$ **c** $4(3 - 2p)$ **d** $4(5 - 4t)$

8 **a** $2(2t + 3)$ **b** $3(2x + 3)$ **c** $2(4t + 3)$ **d** $3(3x + 2)$
e $3(3x - 1)$ **f** $5(2t + 1)$ **g** $4(2x + 1)$ **h** $3(4t + 3)$
i $4(3t + 2)$ **j** $2(4x + 1)$ **k** $3(5t + 4)$ **l** $8(3x - 2)$

9 **a** $2(8x + 5)$ **b** $7(2x - 1)$ **c** $5(3y + 5)$ **d** $5(2y - 1)$
e $3(5m - 6)$ **f** $4(2t + 5)$ **g** $4(3t - 2)$ **h** $4(3 + 4k)$
i $2(5 - 6y)$ **j** $6(5 - m)$ **k** $5(7 + 2k)$ **l** $7(3q + 2)$

10 **a** She hasn't used the HCF.
b It is a correct factorisation but it isn't fully factorised.

Investigation: Interesting numbers

A–D Answers will vary depending on the numbers chosen.

E 22

F **a** pupils' own work
b The answer is always 22.

Lesson 2.3 Equations with brackets

Learning objective
- To solve equations with one or more sets of brackets

Resources and homework
- Pupil Book 3.2, pages 32–34
- Homework Book 3, section 2.3
- Online homework 2.3, questions 1–10

Links to other subjects
- **Science** – to solve equations in physics
- **Business studies** – to do financial calculations

Key words
- No new key words for this topic

Problem solving and reasoning help
- The **PS** questions in Exercise 2C of the Pupil Book require pupils to formulate and solve equations from 2D shapes with unknown side lengths. Ensure that pupils can work out the perimeter and area of 2D shapes before they attempt questions 4, 5 and 6. **Less able** pupils may find it difficult to set up equations. It may be helpful to use values for the unknowns, showing the structure of each equation.

Common misconceptions and remediation
- A common problem often seen when expanding a bracket is to multiply the first term by the number outside the bracket and write down only the second term. Occasionally, pupils may add instead of multiply. Use Part 1 of this lesson to ensure that pupils can expand brackets correctly before tackling Exercise 2C.

Probing questions
- Show pupils a list of different linear equations with brackets and ask:
  - Which can be solved easily?
  - Which are more difficult to solve and why?
  - What strategies should you use with the more difficult ones and why?

Part 1
- Ask pupils to expand the brackets in these expressions:

  $4(x + 4)$ $2(2x - 5)$ $3(3x + 4)$ $\frac{1}{3}(9x - 30)$
- Now show the class examples of errors when multiplying out brackets with errors. Ask pupils to explain what the errors are in each of these expressions:

  $3(x + 4) = 3x + 4$ $2(x - 4) = 2x - 6$ $-2(6 - 2x) = -12 - 4x$

Part 2
- Write the equation $3(3x + 1) = 21$ on the board. Ask pupils how they might start to solve it.
- If pupils suggest multiplying out the bracket first, then write this and the rest of the solution on the board ($x = 2$), asking the class for the next step at each stage.
- Now ask if there was a different way to start the solution. The alternative that you are looking for is to start by dividing both sides by 3 to give $3x + 1 = 7$.
- Work through the solution to obtain the same answer as with the previous method.
- **Pupils can now do Exercise 2C from Pupil Book 3.1.**

Part 3

- Ask if anyone can remember the name of the type of equation they have been looking at today. (Linear equations) Explain that this sort of equation crops up in many different areas of mathematics, science and technology. Being able to solve equations like this (and more complicated equations) is essential for being able to solve problems in these subjects.
- Try to get across the idea that solving equations is like solving a puzzle, but that when solving equations we are trying to use a logical sequence to help us solve the puzzle. The logical sequence is guaranteed to give the correct answer every time, if we do it correctly.

Answers

Exercise 2C

1 **a** $m = 4$ **b** $y = 12$ **c** $k = -4$ **d** $n = 15$
e $k = 8$ **f** $x = 11$ **g** $y = 8$ **h** $t = 9$

2 **a** $t = 3$ **b** $t = 10$ **c** $m = 9$ **d** $x = 0$
e $y = 1$ **f** $p = 10$ **g** $t = 6$ **h** $k = 7$
i $q = 8$ **j** $t = 1$ **k** $m = 11$ **l** $g = 10$
m $t = 15$ **n** $n = 11$ **o** $y = 2$ **p** $q = 14$

3 **a** $m = -1$ **b** $t = -2$ **c** $n = -5$ **d** $q = 3$
e $t = -2$ **f** $k = -2$ **g** $p = -5$ **h** $t = -2$
i $a = -1$ **j** $t = -1$ **k** $h = -1$ **l** $p = 4$
m $d = -4$ **n** $x = -4$ **o** $t = -2$ **p** $m = -4$

4 **a i** $2(x + 5) = 24$ **ii** $x = 7$ **b i** $2(x + 5) = 36$ **ii** $x = 13$
c i $5x = 75$ **ii** $x = 15$ **d** 42 cm

5 **a i** $5(8 + t) = 85$ **ii** $t = 9$ **b i** $5(8 + t) = 90$ **ii** $t = 10$
c i $5(8 + t) = 15$ **ii** $t = 7$ **d** $t = 12$

6 **a i** $180(n - 2) = 180$ **ii** $n = 3$ **iii** triangle
b i $180(n - 2) = 1080$ **ii** $n = 8$ **iii** octagon
c i $180(n - 2) = 1260$ **ii** $n = 9$ **iii** nonagon
d 12

Challenge: Primes from primes

A Yes, all prime numbers greater than 2 are odd so adding them together will give an even number. No even number greater than 2 is prime.

B yes, for example $3 + 5 + 11 = 19$

Lesson 2.4 Equations with fractions

Learning objective

- To solve equations involving fractions

Resources and homework

- Pupil Book 3.1, pages 34–36
- Intervention Workbook 3, page 15–17
- Homework Book 3, section 2.4
- Online homework 2.4, questions 1–10

Links to other subjects

- **Business studies** – to solve problems involving money

Key words

- No new key words for this topic

Problem solving and reasoning help

- Questions 6 and 7 in Exercise 2D of the Pupil Book require pupils to formulate and solve equations. **Less able** pupils often find it difficult to set up equations, so it might be helpful to use values for the unknowns, showing the structure of each equation. These pupils may also struggle with worded questions. Model some examples of worded questions, picking out the key words from each question before pupils attempt to answer them.

Common misconceptions and remediation

- When solving equations that contain fractions pupils sometimes remove the fraction by multiplying, but leave the denominator next to the bracket, for example, pupils may write $\frac{1}{4}(x + 2) = 5$ *incorrectly* as: $4(x + 2) = 20$.

Probing questions

- Show pupils a list of different linear equations involving brackets and fractions and ask:
    - Which can be solved easily?
    - Which are more difficult to solve and why?
- What strategies should you use with the more difficult equations and why would they make solving the equations easier?

Part 1

- Ask for some equivalent fractions to $\frac{1}{2}$. After some correct suggestions of $\frac{2}{4}$, $\frac{5}{10}$, …, ask for an equivalent fraction to $\frac{1}{2}$ that uses the number 34. Two possible answers are $\frac{34}{68}$ and $\frac{17}{34}$.
- Now ask for two fractions equivalent to $\frac{1}{3}$ that use the number 12 ($\frac{12}{36}$ and $\frac{4}{12}$).
- Repeat with the following examples: equivalent to $\frac{1}{5}$ using the number 45 ($\frac{45}{225}$ and $\frac{9}{45}$); equivalent to $\frac{2}{3}$ using the number 18 ($\frac{18}{27}$ and $\frac{12}{18}$); equivalent to $\frac{1}{4}$ using the number 28 ($\frac{28}{112}$ and $\frac{7}{28}$); equivalent to $\frac{3}{4}$ using the number 36 ($\frac{36}{48}$ and $\frac{27}{36}$).

Part 2

- On the board, write the equation $\frac{x}{3} = 5$. Ask the class how to solve it.
- Pupils should respond with: 'multiply both sides by 3', which gives $x = 15$.
- Explain that this is a simple equation using fractions, but they did meet a more difficult equation during the last lesson, for example: $\frac{(4x+5)}{3} = 7$. Ask, 'how do we solve this?'
- Say that, again, it is a matter of simplifying the side around the variable (x) step by step.

- Start by multiplying both sides by 3 to give $4x + 5 = 21$, then subtract 5 from both sides to give $4x = 16$, then divide both sides by 4 to give: $x = 4$.
- Now show the class the following example: $\frac{1}{4}(2x - 6) = 8$
  - Start by multiplying both sides by 4 to give $2x - 6 = 32$.
  - Next, add 6 to both sides to give $2x = 38$.
  - Finally divide both sides by 2, giving $x = 19$
- **Pupils can now do Exercise 2D from Pupil Book 3.1.**

Part 3

- On the board, write the equation $\frac{12}{x} = 2$. Ask the class if anyone can solve the problem.
- Pupils should see that the solution is $x = 6$, but discuss the methods that pupils might have used in order to get that solution if it had been a much more difficult problem.
- Give the class the problem $\frac{245}{x} = 14$ and ask how this could be solved. (Multiply both sides by x to give $245 = 14x$, then divide both sides by 14 to give $x = 17.5$.)

Answers

Exercise 2D

1 **a** $t = 5$ **b** $m = 4$ **c** $y = 4$ **d** $p = 11$ **e** $x = 5$ **f** $q = 4$
g $n = 8$ **h** $a = 3$ **i** $h = 12$ **j** $n = 6$ **k** $x = 4$ **l** $q = 9$

2 **a** $x = 20$ **b** $x = 27$ **c** $x = 12$ **d** $x = 25$
e $x = 8$ **f** $x = 8$ **g** $x = 21$ **h** $x = 20$

3 **a** $t = 7$ **b** $x = 8$ **c** $m = 10$ **d** $x = 7$
e $k = -1$ **f** $t = 17$ **g** $x = 9$ **h** $y = 28$

4 **a** $t = 7$ **b** $x = 8$ **c** $m = 7$ **d** $t = 6$
e $k = 3$ **f** $t = 45$ **g** $x = 5$ **h** $y = 19$

5 **d** is the odd one out with a solution of $x = -6$. All the rest have the solution $x = 6$.

6 **a** $\frac{n}{3}$ **b** $\frac{n}{3} = 7$ **c** $n = 21$

7 **a** $\frac{n+2}{5}$ **b** $\frac{n+2}{5} = 3$ **c** $n = 13$ **d** 28

8 Multiplying by $\frac{1}{5}$ and dividing by 5 are the same. In both cases the solution is $x = -3$.

Mathematical reasoning: Making equations

A pupils' own checks

B **a** $x = 9$
$x - 5 = 4$
$2(x - 5) = 8$
$\frac{2(x-5)}{4} = 2$

b pupils' own checks

C pupils' own equations

Lesson 2.5 Formulae

Learning objective

- To practise using formulae

Resources and homework

- Pupil Book 3.1, pages 37–39
- Intervention Workbook 2, pages 36–39
- Intervention Workbook 3, pages 34–36
- Homework Book 3, section 2.5
- Online homework 2.5, questions 1–10

Links to other subjects

- **Science** – to use equations such as Ohm's Law, mass, density, volume
- **Business studies** – to do financial calculations

Key words

- formula
- variable
- subject

Problem solving and reasoning help

- The activity at the end of Exercise 2E of the Pupil Book enables pupils to demonstrate their learning. **Less able** pupils may need you to demonstrate how to use the flowchart; a quick review of substituting into expressions might also help. Ask **more able** pupils to design their own flowcharts from expressions.

Common misconceptions and remediation

- Pupils sometimes struggle with the concept that letters can represent numbers and that variables within a formula can change. Explain clearly and model some examples so that pupils understand these concepts before tackling any questions.

Probing questions

- When you substitute $x = 3$ and $y = 4$ into the formula $z = xy + 4x$ you get 24. Can you make up some more formulae that also give $z = 24$ when $x = 2$ and $y = 8$ are substituted?

Part 1

- Ask pupils how much text messages cost. Make a list of the different rates (at the time of printing: 8p, 10p or 12p). Use pupils' suggestions, unless they suggest only 10p.
- Ask how many text messages pupils send per day. Using the price per message, work out how much various pupils spend on text messaging per day.
- Ask other pupils how many messages they send per week. Again, using the price per message, work out how much various pupils spend on text messaging per week.
- Talk about top-up cards costing £10, £15 or £20. How many text messages can pupils send from one top up at the different values?
- This is a rich source of mental work that works best when the real live data is being used, that is, current charges. Try not to be exclusive. If some pupils are unable to text, ensure that the wording you use is such that these pupils are included in the discussions.
- You could expand this activity into estimating how many text messages each pupil might make in a year and therefore how much this will cost.

Part 2

- Introduce the topic of changing the subject of a formula by asking pupils how often they need to change their clothes. Bring the class round to formulae and the fact that as much as we need to change each day, so formulae also need to change, depending on what they need to do.
- On the board, write the formula $C = 250 + 5W$. Explain that this formula is used to calculate the cost for advertisements in a certain newspaper, where C is the cost in pence of the advertisement and W is the number of words in the advertisement. Explain that C is the *subject of the formula* because it is the variable (letter) in the formula, which stands on its own, usually on the left-hand side of the equals sign.
- Ask how much it would cost to place a 20-word advertisement in the newspaper.
- Use this example to verify that all pupils can substitute $W = 20$ into the formula to get $C = 250 + 5 \times 20 = 250 + 100 = 350$, giving the cost as £3.50.
- Ask pupils if anyone can think of a formula that uses two variables to work out a third, for example, the 'area of a triangle' formula. If there are no other suggestions, use the triangle formula. Illustrate how to substitute into the formula to calculate one value from two others.
- The area of a triangle is given by $A = \frac{1}{2}bh$, where b is the base length and h is the vertical height of the triangle. The area of a triangle with a base length of 5 cm and a vertical height of 8 cm, is given by: $A = \frac{1}{2} \times 5 \times 8 = 20 \text{ cm}^2$
- **Pupils can now do Exercise 2E from Pupil Book 3.1.**

Part 3

- Ask pupils for as many formulae to do with perimeter, area, surface area and volume as they can think of. Write these on the board. Now give pupils values to substitute into them.

Answers

Exercise 2E

1 **a** 12 cm **b** 6 cm **c** 39 cm
2 **a** 540° **b** 720°
3 **a** £44 **b** £70
4 **a** £30 **b** £43
5 **a** £17 000 **b** £16 000 **c** £10 000
6 **a** 200 m/s **b** 330 m/s
7 **a** 10 cm^2 **b** 42 m^2
8 **a** £270 **b** £210

Activity: Using a flowchart

A pupils' own checks
B The next 8 values are 6.6875, 6.265625, 5.94921875, 5.711914063, 5.533935547, 5.40045166, 5.300338745 and 5.225254059.
C The answer is getting closer to 5 each time.
D no

Review questions

(Pupil Book pages 40–41)

- The review questions will help to determine pupils' abilities with regard to the material within Chapter 2.
- These questions also draw on the mathematics covered in earlier chapters of the book to encourage pupils to make links between different topics.
- The answers are on the next page of this Teacher Pack.

Financial skills – Wedding day

(Pupil Book pages 42–43)

- With the class, go through the information given in the financial skills wedding day activity on pages 42 and 43 of the Pupil Book.
- Explain to pupils how the church cost formula was calculated: Cost = £100 + £90 × number of hours open.
- Ask pupils to calculate the costs for 2, 3 and 4 hours. (£280, £370, £460)
- Explain to pupils how the cars cost formula was calculated: Cost = £150 + £8 × number of miles driven.
- Ask pupils to calculate the costs for 10, 15 and 20 miles driven. (£230, £270, £310)
- Explain to pupils how the formula for the wedding venue was calculated: Cost = £140 + number of hours × £60.
- Ask pupils to calculate the costs for 4, 5 and 8 hours. (£380, £440, £620)
- You could also ask pupils questions such as:
  - What does the £100 stand for in the church formula? How much would it cost if you booked it but did not use it?
  - What is the significance of the £8 in the cars formula?
- Pupils can now do questions 1 to 5 of the activity.

Answers to Review questions

1	**a** $8p$	**b** $9x$	**c** $11q$	**d** $11t$
	e $12n$	**f** $9p$	**g** $6m$	**h** $6a$
	i $7h$	**j** $3g$	**k** $3n$	**l** $5t$
2	**a** $P = 3t + 8$	**b** $P = 4x + 9$	**c** $P = 11m + 8$	
3	**a** 3, 5, 15	**b** 3, 15, 18, 9, 36	**c** 25, 4, 9, 36, 49	
	d 2, 3, 5, 7, 31	**e** 2, 3, 5, 10, 15, 4	**f** 2, 5, 10, 25, 4	
4	**a** $t = 4$	**b** $x = 16$	**c** $m = 7$	**d** $p = 13$
	e $t = 10$	**f** $y = 8$	**g** $n = 6$	**h** $q = 11$
5	**a** $m = 6$	**b** $p = 6$	**c** $x = 5$	**d** $q = 9$
	e $x = 12$	**f** $x = 20$	**g** $x = 27$	**h** $x = 24$
6	**a** $3m + 15$	**b** $3t + 21$	**c** $2x - 10$	**d** $4t - 12$
	e $14 + 2y$	**f** $12 - 4h$	**g** $20 - 4t$	**h** $10 - 5t$
7	**a** $2(3t + 2)$	**b** $3(3x + 4)$	**c** $4(t + 2)$	**d** $4(3x + 2)$
	e $3(3x - 2)$	**f** $5(3t + 1)$	**g** $2(x + 3)$	**h** $7(2t + 1)$
	i $6(3t + 2)$	**j** $2(5x + 3)$	**k** $4(4t + 3)$	**l** $3(2x - 3)$
8	**a** 3	**b** $4t + 3$	**c** $5(2 - 3t)$	
9	**a** $m = 1$	**b** $t = 15$	**c** $x = 4$	**d** $y = 27$
	e $n = 2$	**f** $k = 23$	**g** $x = -1$	**h** $t = 23$
10	**a** $\frac{n+3}{4}$	**b** $\frac{n+3}{4} = 2$	**c** $n = 5$	**d** $n = 25$
11	**a** £25	**b** £38.50		

Answers to Financial skills – Wedding day

1 £505
2 £764
3 £17 970
4 £3395
5 £22 634

3 Polygons

Learning objectives

- The names of different polygons
- The difference between an irregular polygon and a regular polygon
- How to work out the sum of the interior angles of a polygon
- How to work out the size of each interior angle in regular polygons

Prior knowledge

- The different names for triangles and quadrilaterals
- That the sum of the interior angles in a triangle is 180°
- That the sum of the interior angles in a quadrilateral is 360°

Context

- This chapter starts by building pupils' ability to categorise using polygons. This is a good transferable skill across mathematics and beyond. It then introduces pupils to finding the sums of the interior and exterior angles of polygons.

Discussion points

- One of the lines of symmetry of a regular polygon goes through two vertices of the polygon. Convince me that the polygon must have an even number of sides.
- Sketch a shape that will help to convince me that:
    - a trapezium might not be a parallelogram
    - a trapezium might not have a line of symmetry
    - every parallelogram is also a trapezium.

Associated Collins ICT resources

- Chapter 3 interactive activities on Collins Connect online platform
- *Calculating interior and exterior angles of a polygon* video on Collins Connect online platform
- *Church floor* and *La Seta* Wonder of Maths on Collins Connect online platform

Curriculum references

Develop fluency

- Use language and properties precisely to analyse 2D and 3D shapes

Geometry and measures

- Identify properties of, and describe the results of, translations, rotations and reflections applied to given figures
- Derive and use the sum of angles in a triangle and use it to deduce the angle sum in any polygon, and to derive properties of regular polygons

Fast-track for classes following a 2-year scheme of work

- Lesson 3.1 should be familiar material for pupils. Check pupils' knowledge by giving them some questions. If all pupils can answer them and you are satisfied that everyone in the class understands the material, then move on to Lesson 3.2.

Lesson 3.1 Polygons

Learning objectives

- To know the names of polygons
- To know the difference between an irregular polygon and a regular polygon

Links to other subjects

- **Design and technology** – to construct models of designs
- **Geography** – to identify locations on maps by descriptions of the 2D representation of man-made objects

Resources and homework

- Pupil Book 3.1, pages 45–48
- Intervention Workbook 2, pages 46–48
- Homework Book 3, section 3.1
- Online homework 3.1, questions 1–10

Key words

- concave polygon
- convex polygon
- decagon
- heptagon
- hexagon
- irregular polygon
- nonagon
- octagon
- pentagon
- polygon
- regular polygon

Problem solving and reasoning help

- The investigation at the end of Exercise 3A in the Pupil Book requires pupils to apply their learning to an abstract multi-step problem. Use guided group work to support pupils who struggle to apply what they have learnt independently, or challenge pupils to develop their explanations if appropriate.

Common misconceptions and remediation

- The ability to categorise is a valuable transferable skill across mathematics and beyond. Help pupils to see how they could apply this skill to solving problems more generally.

Probing questions

- Sketch me a quadrilateral that has, for example, one line of symmetry, two lines, three lines, no lines, and so on. Can you give me any others? What is the order of rotational symmetry of each of the quadrilaterals you sketched?
- One of the lines of symmetry of a regular polygon goes through two vertices of the polygon. Convince me that the polygon must have an even number of sides.

Part 1

- Draw a set of polygons on the board. Working in pairs, ask pupils to write down as much as they can about each polygon. Then take feedback.
- You could also approach this as an 'odd one out' activity.

Part 2

- Ask pupils to summarise the types of information they used in Part 1, for example: names, number of sides, angles, regular or irregular, convex or concave.
- Use pages 45 and 46 of the Pupil Book for support, as needed.

Part 3

- While working in pairs to consolidate and extend their learning, ask pupils to provide examples for the following question: sketch a shape to show that:
  - a trapezium might not be a parallelogram
  - a trapezium might not have a line of symmetry
  - every parallelogram is also a trapezium.

Answers

Exercise 3A

1 **a** hexagon **b** decagon **c** pentagon
d octagon **e** nonagon **f** heptagon

2 **a** yes, pentagon **b** no **c** yes, octagon
d no **e** yes, heptagon

3 **a** no **b** yes **c** yes **d** yes **e** no

4 **a** convex **b** convex **c** concave **d** concave **c** convex

5 **a** for example **b** for example

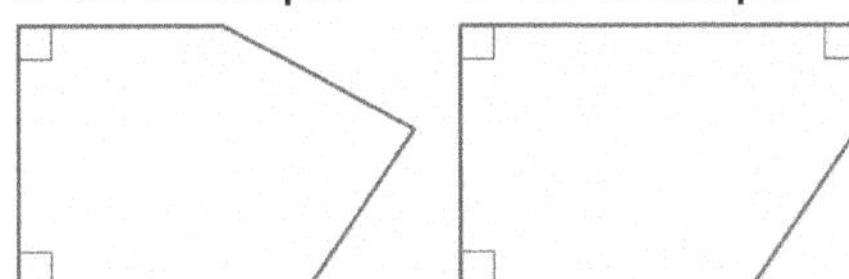

c You would end up with a rectangle or a square.

6 **a** AB is parallel to ED and AF is parallel to CD and BC is parallel to FE.
b AB is parallel to FE and BC is parallel to GF and CD is parallel to HG and DE is parallel to AH.

Investigation: Overlapping squares

A Polygon A is a quadrilateral, polygon B is a triangle and polygon C is a hexagon.

B Polygon A is a hexagon, polygon B is a quadrilateral and polygon C is a hexagon.

C for example: triangle, quadrilateral, pentagon, hexagon and heptagon

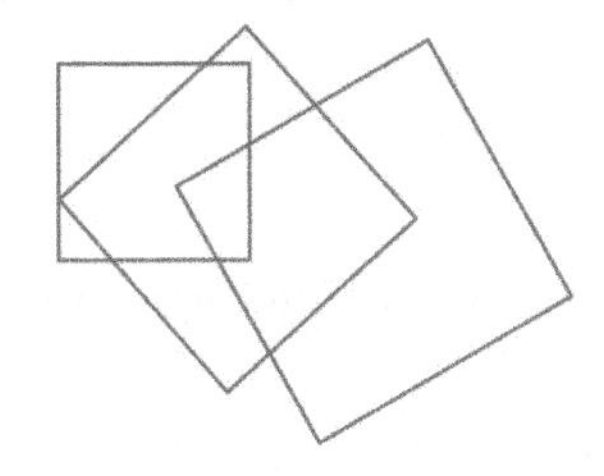

Lesson 3.2 Angles in polygons

Learning objective

- To work out the sum of the interior angles of a polygon

Resources and homework

- Pupil Book 3.1, pages 48–51
- Intervention Workbook 3, pages 52–53
- Homework Book 3, section 3.1
- Online homework 3.2, questions 1–10

Links to other subjects

- **Design and technology** – to construct models of designs

Key words

- interior angle

Problem solving and reasoning help

- **PS** question 7 of Exercise 3B requires pupils to apply their understanding of algebra to generalise their learning from this lesson. This is extended in the problem solving activity at the end of the exercise.

Common misconceptions and remediation

- Pupils try to learn justifications of this type by memory instead of understanding the logic and therefore the transferable nature of the thinking process. Discuss this with pupils so that it is clear in their minds. Then give them the opportunity to identify the steps in the process and use similar logic in different contexts.

Probing questions

- What clues do you look for when solving a geometrical problem? How do you decide where to start? Is it possible to solve the problem in a different way?
- How would you convince a friend that the exterior angles of a polygon add up to 360°?

Part 1

- Ask pupils to imagine an equilateral triangle ABC. Then ask pupils to imagine that the base line AB is fixed and the vertex C is allowed to move parallel to the base.
- Next, ask pupils to describe what other types of triangles can be formed as C moves. (scalene triangles, two right-angled triangles and two isosceles triangles)

Part 2

- Remind the class that the sum of the interior angles of a triangle is 180°, and that the sum of the interior angles of a quadrilateral is 360°. Illustrate these points by splitting a quadrilateral into two triangles.
- Ask the class to write down the names of all the other *polygons* that they met in Year 7: pentagon, hexagon, heptagon, octagon, nonagon, decagon.
- Make sure that pupils understand the definition of a *regular polygon*.
- Show the class how to find the sum of the *interior angles* of a hexagon. Show how a hexagon can be split into four triangles from one of its vertices.

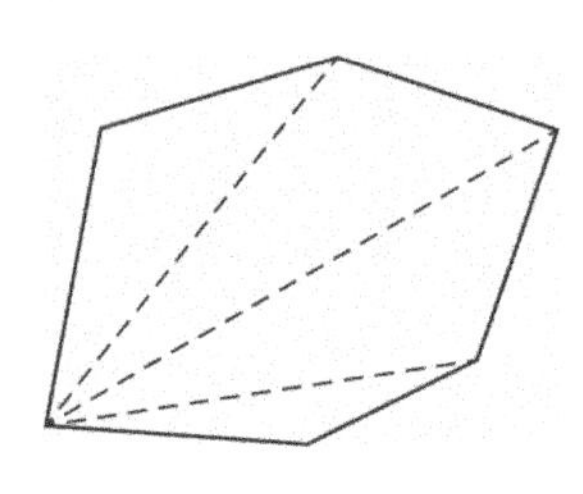

- The sum of the interior angles for each triangle is 180°. So, the sum of the interior angles of a hexagon is 4 × 180° = 720°. Tell pupils they will need to apply this idea to the activities in Exercise 3B.
- **Pupils can now do Exercise 3B from Pupil Book 3.1**.

Part 3

- Invite a pupil to the board to explain how to find the sum of the interior angles of a polygon.
- Ask another pupil to give the value of the sum of the exterior angles of a polygon.
- Invite a third pupil to explain, on the board, the connection between the interior angle and the exterior angle at any vertex of a polygon.

Answers

Exercise 3B

1 A hexagon can be split into 4 triangles.
So the sum of the interior angles of a hexagon is given by: $4 \times 180° = 720°$

2 A heptagon can be split into 5 triangles.
So the sum of the interior angles of a heptagon is given by: $5 \times 180° = 900°$

3 An octagon can be split into 6 triangles.
So the sum of the interior angles of an octagon is given by: $6 \times 180° = 1080°$

4

Name of polygon	Number of sides	Number of triangles inside polygon	Sum of interior angles
Triangle	3	1	180°
Quadrilateral	4	2	360°
Pentagon	5	3	540°
Hexagon	6	4	720°
Heptagon	7	5	900°
Octagon	8	6	1080°
Nonagon	9	7	1260°
Decagon	10	8	1440°

5 The sum of the interior angles of a hexagon is 720°.
So $a = 720° - 150° - 70° - 140° - 130° - 120° = 110°$.

6 **a** 100° **b** 125 **c** 290°

7 120

Problem solving: Polygons and diagonals

A

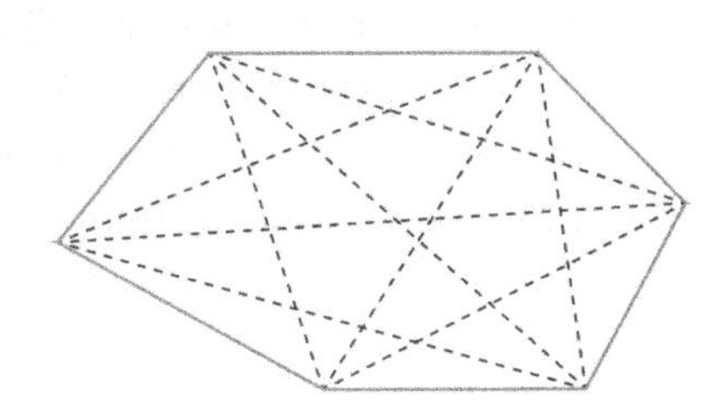

B **a** $d = \frac{7(7-3)}{2} = \frac{7 \times 4}{2} = \frac{28}{2} = 14$ **b** $d = \frac{8(8-3)}{2} = \frac{8 \times 5}{2} = \frac{40}{2} = 20$

C

Number of sides	Name of polygon	Number of diagonals
3	triangle	0
4	quadrilateral	2
5	pentagon	5
6	hexagon	9
7	heptagon	14
8	octagon	20
9	nonagon	27
10	decagon	35

Lesson 3.3 Interior angles of regular polygons

Learning objective

- To work out the sizes of interior angles in regular polygons

Resources and homework

- Pupil Book 3.1, pages 52–53
- Homework Book 3, section 3.2
- Online homework 3.3, questions 1–10

Links to other subjects

- **Design and technology** – to construct models of designs

Key words

- No new key words for this topic

Problem solving and reasoning help

- **PS** question 5 of Exercise 3C in the Pupil Book requires pupils to use algebra to come to a solution and **PS** question 6 requires them to summarise what they have learnt using a less familiar polygon. The challenge at the end of the exercise requires pupils to draw on prior learning from different areas of mathematics – in this case, decimals and fractions.

Common misconceptions and remediation

- Pupils try to learn justifications of this type by memory instead of understanding the logic and therefore the transferable nature of the thinking process. Discuss this with pupils so that they are clear on this. Provide opportunities for pupils to identify the steps in the process and to use similar logic in different contexts.

Probing questions

- What clues do you look for when solving a geometrical problem? How do you decide where to start? Is it possible to solve the problem in a different way?
- How would you convince a friend that each angle in a ……. is equal to …….?

Part 1

- Ask pupils to imagine a square, then to imagine cutting the square along one diagonal. Then ask pupils to describe the two shapes that are left. (two isosceles right-angled triangles)
- Ask pupils to imagine cutting the square again, but this time along a line that is parallel to the diagonal. Finally, ask pupils to describe the two shapes that are left. (an isosceles right-angled triangle and a pentagon)

Part 2

- Remind the class of the definition of a regular polygon. A polygon is regular when all its interior angles are equal and all its sides have the same length.
- Explain that this lesson is about how to calculate the size of each exterior and interior angle of any regular polygon.
- Draw a regular pentagon on the board or OHP, with one of its exterior angles labelled x and the interior angle labelled y, as shown in the diagram.

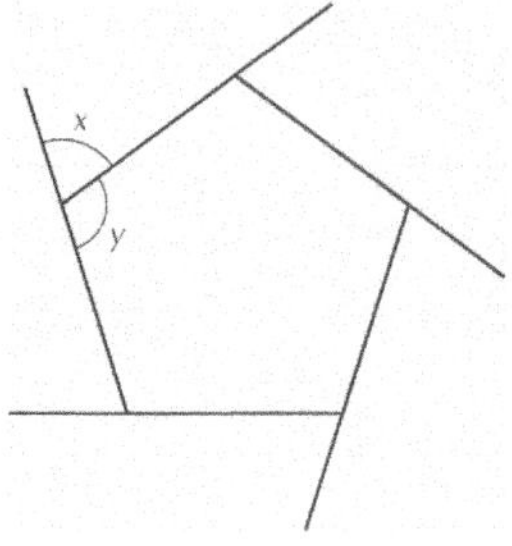

- The regular pentagon has five equal exterior angles, the sum of which is 360°. So, we have: $5x = 360°$; $x = 72°$.
- The regular pentagon also has five equal interior angles. The sum of an interior angle and exterior angle is 180°. So, we have: $y = 180° - 72° = 108°$.
- **Pupils can now do Exercise 3C from Pupil Book 3.1**.

Part 3

- Conduct a quick revision test to ensure that pupils can recall or calculate the size of the interior angles for common regular polygons.
- Ask pupils to write down the size of each interior angle for each shape:
  1 equilateral triangle **2** square **3** regular pentagon **4** regular hexagon

Answers

Exercise 3C

1 **a** 720° **b** 120°

2 **a** 1080° **b** 135°

3

Name of polygon	Number of sides	Sum of interior angles	Size of each interior angle
Triangle	3	180°	60°
Quadrilateral	4	360°	90°
Pentagon	5	540°	108°
Hexagon	6	720°	120°
Octagon	8	1080°	135°
Nonagon	9	1260°	140°
Decagon	10	1440°	144°

4 **a** 108° **b** isosceles **c** 36° **d** 36° **e** 36°

5 30°

6 **a** 10 **b** 10 × 180° = 1800° **c** 1800° ÷ 12 = 150°

Challenge: Interior angles in a regular heptagon

$128\frac{4}{7}°$

Review questions

(Pupil Book pages 54–55)

- The review questions will help to determine pupils' abilities with regard to the material within Chapter 3.
- These questions also draw on the mathematics covered in earlier chapters of the book to encourage pupils to make links between different topics.
- The answers are on the next page of this Teacher Pack.

Activity – Regular polygons and tessellations

(Pupil Book pages 56–57)

- This activity is designed to give pupils the opportunity to apply what they have learnt about the characteristics of polygons to tessellations. Pupils will need to apply what they know about angles in polygons.
- Tessellations were not part of this chapter but pupils should have met the concept before. You could warm up for this activity by asking pupils what they already know about tessellations and how this might link to what they learnt in this chapter.
- As an extension, pupils could use the internet to investigate the use of tessellating polygons in art and architecture.

Answers to Review questions

1 **a i** hexagon **ii** pentagon **iii** decagon **iv** octagon
b i convex **ii** concave **iii** convex **iv** concave

2 **c** is the odd one out as it is the only regular hexagon.

3 **a** no **b** yes **c** yes **d** yes **e** no

4 **a ii** 720° **iii** 900° **iv** 1080°
b i 140° **ii** 100° **iii** 130° **iv** 117°

5 67.5°

6 36°, 72°, 108°, 144°, 180°

Answers to Activity – Regular polygons and tessellations

1 no

2 yes

3 no

4 yes, yes, no, yes, no

4 Using data

Learning objectives

- How to recognise correlation from scatter graphs
- How to construct and interpret two-way tables
- How to compare two sets of data from statistical diagrams
- How to plan a statistical investigation

Prior knowledge

- How to calculate averages
- How to use a suitable method to collect data
- How to draw and interpret graphs for discrete data
- How to use mode, median, mean and range to compare two sets of data

Context

- This chapter picks up the ideas from the material that pupils learned in statistics in previous years. It develops ways to illustrate distributions and how we can use data to explore possibilities as well as to compare them. The chapter culminates in pupils conducting their own investigations, using the ideas from the first part of the chapter. The following link to a video demonstrates an example of the power of statistics:
  http://www.gapminder.org/videos/the-river-of-myths/

Discussion points

- If two set of data correlate, does this mean that a change in one set causes a change in the other set? Explain why or why not.
- Show pupils a range of different statistical graphs and ask: 'What conclusions can be drawn from this graph? Explain your answer.'
- Which graphs are more difficult than others to interpret? Explain why.

Associated Collins ICT resources

- Chapter 4 interactive activities on Collins Connect online platform

Curriculum references

Reason mathematically

- Explore what can and cannot be inferred in statistical and probabilistic settings, and begin to express their arguments formally

Statistics

- Construct and interpret appropriate tables, charts, and diagrams, including frequency tables, bar charts, pie charts, and pictograms for categorical data, and vertical line (or bar) charts for ungrouped and grouped numerical data

- Describe simple mathematical relationships between two variables (bivariate data) in observational and experimental contexts and illustrate using scatter graphs

Fast-track for classes following a 2-year scheme of work

- Much of the material in the lessons of this chapter will be new to pupils. However, Lesson 4.3 and Lesson 4.4 could be combined, but first make sure that pupils have a good grasp of correlation and time series before moving on.

Lesson 4.1 Scatter graphs and correlation

Learning objective

- To infer a correlation from two related scatter graphs

Resources and homework

- Pupil Book 3.1, pages 59–61
- Homework Book 3, section 4.1
- Online homework 4.1, questions 1–10

Links to other subjects

- **Geography** – to make data comparisons for different countries
- **Physical education** – to compare the progress of athletes

Key words

- negative correlation
- positive correlation
- no correlation
- scatter graph

Problem solving and reasoning help

- The **MR** and **PS** questions in Exercise 4A of the Pupil Book test pupils' understanding of correlation. Most pupils will be able to state the nature of correlation from the scatter graph patterns without problems. However, these questions explore the underlying reasons for the correlation. Pupils must have this deeper understanding of correlation before moving on.

Common misconceptions and remediation

- Explain that correlation does not imply causality, or interconnection, which is a common problem area in pupils' interpretation of correlation.

Probing questions

- Describe the type of correlation you would expect from each of these sets of data if you were to plot them on a scatter diagram: height and arm span; hair colour and IQ; hours watching television and time spent exercising.

Part 1

- Copy this table onto the board (without the answers in brackets). Ask pupils for facts about it. For example, blue is the favourite in both classes; three times as many like blue as red in Class 2, twice as many like 'other' in Class 1. Change the numbers or headings and repeat.

Favourite colour	Class 1		Class 2	
Blue	9	(15)	15	(8)
Red	8	(5)	5	(15)
Yellow	5	(0)	6	(4)
Other	8	(10)	4	(3)

- Now ask pupils to draw a blank table with the same headings and number of rows as above. Working in pairs or groups, pupils should try to complete it with these facts:
    - There are 30 pupils in each class. In Class 1, 25 did not pick red; no one picked yellow.
    - In Class 2, three times as many pupils picked red as in Class 1.
    - 10 more pupils picked blue than red in Class 1.
    - In Class 2, only three pupils picked 'other' and twice as many picked blue as yellow.
- Now ask how many picked yellow in Class 2. Repeat the statements if necessary.

Part 2

- Tell pupils that now they will look at pairs of sets of data to see if there are any connections or relationships between them. Ask what happens to the height of children as they get older.

- Sketch a graph on the board of height against age and plot a cross near the origin. Tell the class that this cross represents a very young child. Ask where the cross will be one year later. Plot the cross. Continue until you have a few crosses with positive correlation. Introduce the words *correlation* and *positive correlation*. You could talk about why the crosses do not fall in a perfect straight line, and about strong or weak correlation.
- Now sketch a negative correlation graph on the board and ask pupils to suggest what the labels could be for the axes, for example: 'Value of a car' and 'Age'.
- Discuss the idea of *negative correlation* and then introduce *no correlation*. Summarise by drawing a graph of each type and asking the class what correlation each one has.
- **Pupils can now do Exercise 4A from Pupil Book 3.1.**

Part 3

- Emphasise that there are other types of statistical graphs as well as those already covered. Mention bar charts, frequency diagrams and line graphs.
- Tell pupils that it is important to choose the graph they will use carefully; they must be able to justify their choice. Explain that in the next lesson they will look at creating scatter graphs.

Answers

Exercise 4A

1 **a** positive correlation **b** no correlation

2 **a** positive correlation **b** no correlation

3 **a** negative correlation

b negative correlation

4 **a**

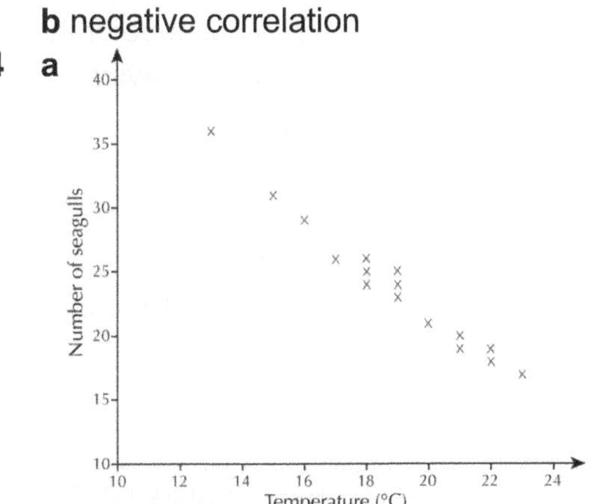

b negative correlation

c Draw a line of best fit through the points on the graph and see what temperature corresponds to 22 gulls.

5

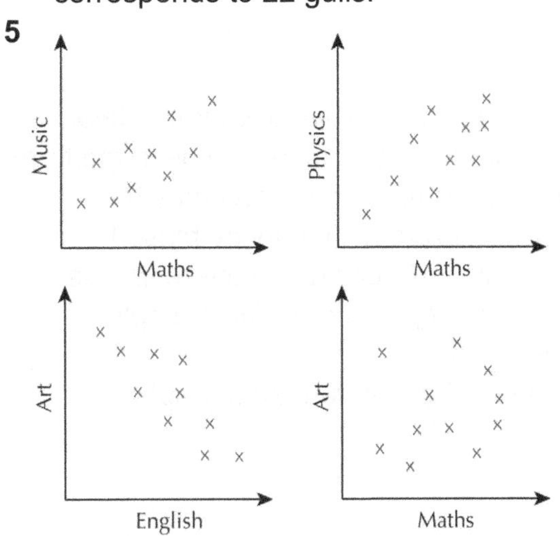

6 **a** **b**

7

8

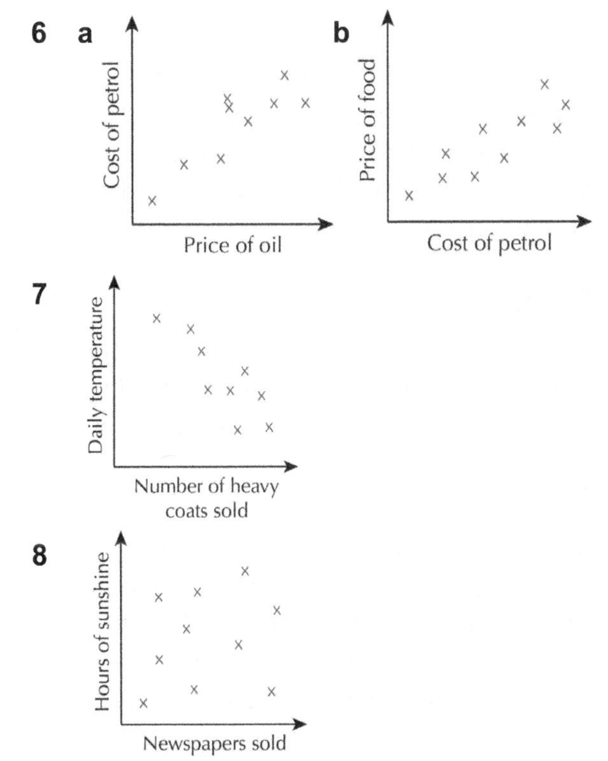

Investigation: Comparing marks

pupils' own answers, for example:
Mathematics/Science positive correlation,
Mathematics/Art positive correlation

Lesson 4.2 Interpreting graphs and diagrams

Learning objective

- To use and interpret a variety of graphs and diagrams

Resources and homework

- Pupil Book 3.1, pages 62–65
- Intervention Workbook 1, pages 60–63
- Intervention Workbook 2, pages 76–79
- Intervention Workbook 3, pages 69–76
- Online homework 4.2, questions 1–10

Links to other subjects

- **Geography** – to compare temperature increase over time
- **History** – to interpret population increase over time

Key words

- No new key words for this topic

Problem solving and reasoning help

- All the questions in Exercise 4B of the Pupil Book are **MR**, and require pupils to interpret a variety of different graphs. Pupils will need to be able to read axes' scales accurately in order to gather correct information for their interpretation of the data.

Common misconceptions and remediation

- Pupils often make lengthy statements but they either fail to support their statements with evidence, or pupils only include some of the evidence. Encourage pupils to bullet point the evidence when making a statement about a graph. Remind pupils that time is always plotted on the horizontal axis.

Probing questions

- Do the intermediate values have any meaning on these line graphs? Why?
- Show graphs where there is no meaning, for example, a line graph showing the trend in temperatures at the same time over a week.

Part 1

- Ask a pupil to come to the front. Give her or him a list of the names of mathematical graphs or charts, for example: pie chart, pictogram, bar chart or frequency diagram.
- Ask the pupil to describe one of the graphs or charts to the class without using its name.
- The class must guess what is being described.
- Extend by using other statistical words such as: correlation, average, questionnaire, survey.

Part 2

- Explain that the aim of the lesson is to interpret graphs and tables using secondary data.
- Discuss the meaning of secondary data and then point out that such data may be in the form of graphs, charts or tables. Explain how data can mislead. Graphs and charts may be deliberately misleading, or they may be accurate but the interpretation may be biased. Ask a pupil how long she or he spent on homework last night. Suggest that the response may vary depending on who asks the question. Answers may be precise: for example: 'I spent 30 minutes'; or imprecise, for example: 'I spent ages.'
- Tell pupils that they will have to criticise how someone else has interpreted a graph.

- Then ask the class to look at Example 1 on page 62 of the Pupil Book. Before reading the question, ask pupils to tell you a fact from the exploded pie chart. Then ask them to look at the complete example and see if they can use different arguments to those in the example.
- **Pupils can now do Exercise 4B from Pupil Book 3.1.**

Part 3

- Ask the class to sketch a circle in their books.
- Tell pupils that you will give some information, which they must use to sketch a pie chart. For example, there are four sectors, say blue, red, green and yellow. The largest sector is more than half the circle. The smallest sector is only slightly smaller than the next sector.
- Now ask the class to complete their pie charts with labels. Point out that they do not have all the required information, so their pie charts will not all be the same.
- Ask the class to compare their pie charts in small groups.
- Summarise the lesson by pointing out that statistics can be misleading when information is missing. Emphasise the need for pupils to include as much detail as possible when writing reports or interpreting graphs and charts.

Answers

Exercise 4B

1 **a** 3 km **b** 30 minutes
c This section of the graph is steeper.
d He has missed the return journey, which is also 8 km.

2 **a** 6 **b** 12
c He could be correct, as the shortest throw is between 0 and 1 metre.
d Yes, the total number of throws is 21.

3 **a** No, just under half of the pie chart represents daffodils.
b You would need to measure the angles for crocuses and anemones.

4 **a** Andy **b** Duncan
c We only know percentages, not the actual number of games.
d David, as he has the lowest percentage of wins and the highest number of losses.
e for example, the number of games may vary greatly between each of the managers; also, the levels of the matches being played could be very different

5 **a** Daisy Down Farm and Bannerdale **b** Dale farm
c Yes, even though two of the farms had decreased in size, the total number of cows had increased by 202.

Challenge: Off their trolley

pupils' own diagrams

Lesson 4.3 Two-way tables

Learning objective

- To interpret a variety of two-way tables

Resources and homework

- Pupil Book 3.1, pages 66–69
- Homework Book 3, section 4.3
- Online homework 4.3, questions 1–10

Links to other subjects

- **Geography** – to read distance tables between places
- **Physical education** – to interpret BMI in two-way tables

Key words

- two-way table

Problem solving and reasoning help

- The questions in Exercise 4C of the Pupil Book encourage pupils to apply their learning in order to deepen their understanding of two-way tables. Help **less able** pupils by modelling several examples with them.

Common misconceptions and remediation

- Pupils do not always read tables accurately by aligning the required row and column. Encourage **less able** pupils to use a ruler on the row and follow the column down with their finger in order to get accurate answers.

Probing questions

- Make up a statement or question for this table using one or more of the following key words: total, range, greatest, smallest, fraction, percentage, probability.

Part 1

- Divide the class into small groups of four or five pupils. Give a mental test of 10 questions. Appoint one person from each group as team captain to record the answers.
- After the answers have been given, ask the team captains to record their answers in a two-way table on the board as shown, using ticks for correct answers.

Group	Q1	Q2	Q3	Q4	Q5	Q6	Q7	Q8	Q9	Q10
1	√		√	√		√			√	√
2	√	√	√		√	√			√	
3		√		√			√	√	√	

- **Test**

1 £4.99 × 4 **2** 25% of 60 **3** $\frac{1}{2}$ of a $\frac{1}{2}$ **4** 600 × 4000

5 72 ÷ 0.2 **6** What is the HCF of 36 and 48?

7 Write down one answer to $x^2 + x = 0$ **8** Increase £132 by 20%

9 What is the square root of 196? **10** Give both solutions to $(5 + x)^2 = 81$

- **Answers**

1 £19.96 **2** 15 **3** $\frac{1}{4}$ **4** 2 400 000

5 360 **6** 12 **7** $x = 0$ or $x = -1$ **8** £158.40

9 14 or –14 **10** $x = 4$ and $x = -14$

Part 2

- Still in their groups, explain to pupils that they need to collect and record data from the class. Use data collection sheets such as those shown below, or let pupils design their own.

- One pupil from each group is the 'collector'; the rest of the group are 'informers'. Each 'collector' should go from group to group collecting the data.

	Favourite subject	
	Boys	**Girls**
English		
Science		
Art		
Maths		

	Favourite colour	
	Boys	**Girls**
Blue		
Red		
Yellow		
Green		

- Now use other combinations to form different two-way tables. For example:

		Favourite colour			
		Blue	**Red**	**Yellow**	**Green**
Favourite hobby	Sport				
	Computer				
	Music				

- After collecting the data, pupils should record it. In each case, ask pupils to pick out a key feature such as: data appears random – no relationship between the two variables.
- **Pupils can now do Exercise 4C from Pupil Book 3.1.**

Part 3

- Ask the class to select a table where they saw a relationship. Look at, for example, boys' favourite colour and boys' favourite music. Are their responses different to those of the girls?
- Write any relationships on the board. Ask the class what they could do to test whether the results were representative of the school.

Answers

Exercise 4C

1 **a** 3 **b** 1 **c** Neil and Paul

2 **a** Philip and Kevin **b** Brian and Malcolm – Brian won **c** Pete and David

3 **a** 3 **b** 28 **c** 15 **d** 7

4 **a** Reikie **b** Jana **c** It is easy to see how many games each person won.
d It allows you to see which games each person won. **e** pupils' own tables

5 **a** As the pupils get older, the number having school lunch decreases.
b Between Y7 and Y8 the reduction is 23, which is the greatest change.
c For example, 68. This reduces the number again but by a smaller amount than between Y8 and Y9.

6 **a** The differences (boys − girls) are: 4%, 3%, 3%, −3%, −3%, −2%.
b At age 10 a higher percentage of boys have android phones, but from age 13 a higher percentage of girls have them.

Activity: A tall story

A

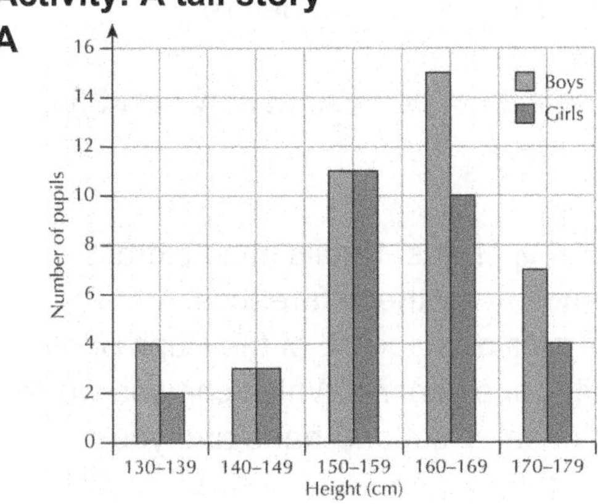

B Yes, as there are more boys in the two tallest categories (although it could be argued that this is just because there are more boys than girls overall).

Lesson 4.4 Comparing two or more sets of data

Learning objective

- To compare two sets of data from statistical tables and diagrams

Resources and homework

- Pupil Book 3.1, pages 70–74
- Intervention Workbook 1, pages 60–63
- Homework Book 3, section 4.4
- Online homework 4.4, questions 1–10

Links to other subjects

- **Science** – to compare experimental data
- **Geography** – to compare data from different countries

Key words

- No new key words for this topic

Problem solving and reasoning help

- All the questions in Exercise 4D of the Pupil Book are **MR** or **PS**. Pupils must interpret data from a variety of different graphs and charts, including pie graphs and composite bar graphs. **Less able** pupils may need some explanation of how to interpret less familiar graphs.

Common misconceptions and remediation

- In order to make comparisons between graphs, pupils need to be able to understand what the graph represents, what the axes mean and how to read data from the graph. Make sure pupils are able to do this before they attempt comparisons. You may need to model some examples for **less able** pupils before they tackle these types of questions.

Probing questions

- What do the axes' labels tell you about the graph?
- What does this part of the graph mean? How can you tell?
- What does this point on the graph tell you?
- Make up some questions about this graph.

Part 1

- Write £20 000 on the board. Explain to pupils that you have seen an advert for a brand new car and £20 000 is the price. Using 'show me' cards, ask pupils to write on their cards what they think the value of the car will be in one year's time.
- Identify the smallest and biggest answers. Ask a pupil to tell you the range of these.
- Ask a pupil to explain why he or she said, for example, £15 000.
- Now repeat, but this time pupils must work in pairs and agree on a value.
- Again look at the range of answers. Hopefully, the value will be less than before.
- Ask a pupil to explain why the range is now less.
- You could repeat this exercise for different items such as a tin of beans that now cost 35p.

Part 2

- Say that the objective of this lesson is to compare graphs or charts. Pupils must extract important information from the graphs or charts and comment on the differences.
- Ask pupils to look at the two graphs about the values of cars on page 70 of the Pupil Book.
- Ask questions such as: 'Which car had a lower value after one year?'; 'Which car was worth only £3000 after four years?'; 'If you were buying car X or car Y second-hand and you wanted the cheaper one, which would you buy?'

- Ask pupils to draw a graph of a car with the price starting at £20 000 and losing £4000 in value every year. Ask a pupil to explain why this is impossible. Prompt the response that after six years the car would have a negative value.
- Now draw a percentage bar chart. Ask how many of pupils walk home after school.
- Mark off the approximate percentage on the bar. Shade it in.
- Now draw another percentage bar on the chart. Tell the class that another group that you teach gave a different result. Mark this result on the bar. Ask pupils to compare the two bars.
- **Pupils can now do Exercise 4D from Pupil Book 3.1.**

Part 3

- Describe two villages to the class. For each village, pupils must sketch a pie chart.
    - Village 1: half the population is aged over 60; one-third of the population is under 18.
    - Village 2: smaller proportion of over 60s than Village 1; larger proportion of under 18s.
- Ask pupils to compare their pie charts with each other to see if they have interpreted the data correctly. There will be slight differences in their answers.

Answers

Exercise 4D

1 **a** 14 **b** winter **c** Yes, as 6 girls chose spring but only 2 boys did.

2 **a** Both plants had 42 tomatoes.
b The blue light had no effect on the overall number of tomatoes.

3 **a** Yes, both schools have the same number of pupils and the cycling sector for Conchord Park is slightly larger than for Bradway school. **b** The sector for 'Walk' is twice the size.
c For example, pupils live closer to the school or the school is not on a bus route.

4 For example, a higher proportion of children attended the brass band than the rock band.

5 The pupils found the reading test harder, as there were less high marks recorded for this test.

6 **a** The highest percentage of buses are late between 9 am and 12 noon. This percentage decreases as the day goes on.
b You would expect the majority of buses to be on time.

7 **a** Peppa Pig **b** The Beano **c** pupils' own diagrams

Activity: How many?

A

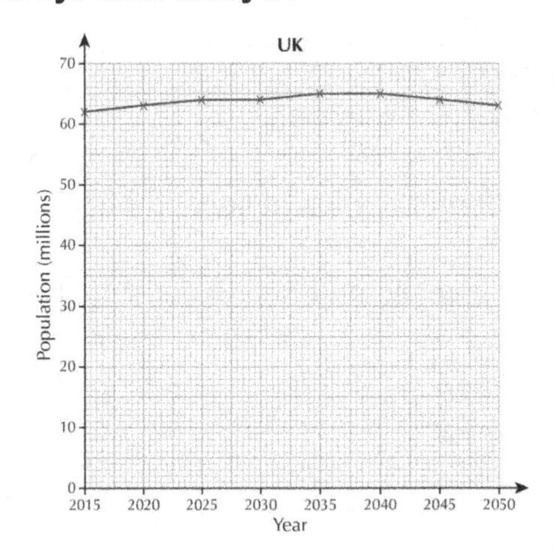

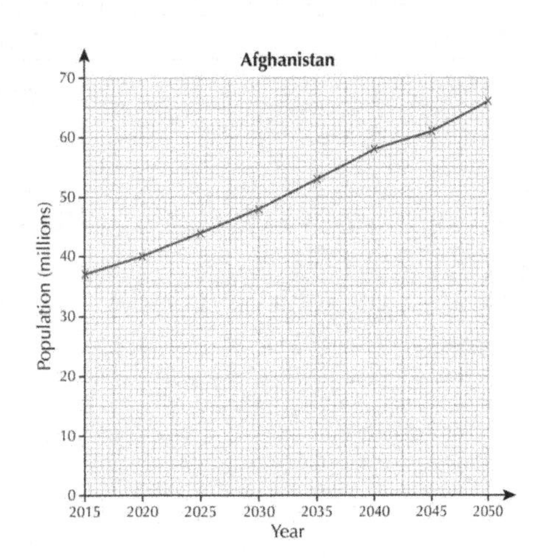

B Around 2048

C UK around 62 million, falling slowly; Afghanistan around 70 million, increasing rapidly

Lesson 4.5 Statistical investigations

Learning objective

- To plan a statistical investigation

Resources and homework

- Pupil Book 3.1, pages 74–77
- Homework Book 3, section 4.5

Links to other subjects

- **Science** – to plan experimental investigations

Key words

- No new key words for this topic

Problem solving and reasoning help

- The investigation at the end of Exercise 4E in the Pupil Book enables pupils to apply their understanding of the skills learned in this lesson to a real-life statistical investigation.

Common misconceptions and remediation

- It is important that pupils' data collection sheets are fit for purpose and have been designed to capture all the factors that have a bearing on the investigation.
- Another common issue is collecting too much data. This has the effect of slowing down the investigation and may confuse **less able** pupils.

Probing questions

- How do you design a good data collection sheet?
- What size of sample is most appropriate?
- What is important with regard to the degree of accuracy in the data you are collecting?
- How will you interpret the data you have collected? How will you display your results?

Part 1

- Pupils can work in small groups. Write the following sources of information on a sheet: Questionnaire, Printed tables in books, Internet, Computer database and Observation sheet.
- Give pupils cards, or a list, containing the following topics:

Primary data	Secondary data
Number of left-handed pupils in the class	Long jump performances in international athletes
TV viewing habits of pupils	Car engine sizes
Reaction times of pupils	Populations of various countries
Are pupils better at catching with their left or right hand?	Football results in Europe
The amounts of pocket money received by males and females in school	Prices of different makes of second-hand cars

- Pupils should discuss each topic and decide how best to investigate each one, using the sources given above.

Part 2

- Explain that the aim of this lesson is to look at how to plan a statistical investigation.
- Point out that sometimes the most difficult part is to decide on a topic to investigate. (Steps 1 and 2 in the table on page 76 of the Pupil Book.) Explain to pupils that to help them they will be given a planning sheet so that they may work systematically through their problem.
- Use an example of your choice or select one from the Pupil Book.
- Discuss the remainder of the table in the Pupil Book with the class.
- **Pupils can now do Exercise 4E from Pupil Book 3.1.**

Part 3

- Use a group's planning sheet to discuss the points listed.
- Ask other groups to contribute points that could be added to the planning sheet.
- Explain to pupils that their homework is to produce an individual plan for a different topic. Pupils could use ideas already used by other groups.

Answers

Exercise 4E

Science plan

1 investigate
2 car
3 engine
4 not
5 books
7 petrol, bias
9 nearest

Geography plan

1 compare
3 housing
4 incomes
5 internet
6 average
8 information
12 sample, mean

Investigation: On your bike

pupils' own work

Review questions (Pupil Book pages 78–79)

- The review questions will help to determine pupils' abilities with regard to the material within Chapter 4.
- These questions also draw on the mathematics covered in earlier chapters of the book to encourage pupils to make links between different topics.
- The answers are on the next page of this Teacher Pack.

Challenge – Rainforest deforestation (Pupil Book pages 80–81)

- Talk to pupils about deforestation and the fact that for years, the big rainforests of the world have been reduced and chopped down, while the country or countries concerned benefit from the cleared land and the revenue from the wood obtained from the trees. Note that this challenge has no intention of making any judgement values of the country or countries concerned. Instead, it has been devised to allow pupils to find what the statistics may suggest; in other words, that economic growth can affect the amount of deforestation.
- Remind the class about simple pie charts that are split into 10 sectors, with each sector showing 10 per cent.
- Draw an example of such a pie chart on the board. Shade one sector and label it 'children', shade four more sectors and label these 'women'; label the remaining sectors 'men'.
- Say that this pie chart shows the proportion of travellers on a busy morning train between two places near pupils' homes. Ask for the percentages shown for different travellers. (male 50%, female 40% and children 10%)
- The data handling expected in this spread is:
  - reading from a table
  - drawing bar charts
  - reading and constructing pie charts
  - interpreting trends from charts and graphs.
- Then ask pupils to answer the questions in the challenge activity in the Pupil Book.
- Close the lesson with a short discussion about why economic growth might affect deforestation. (For example: As the development in a country increases, there is a need to create more factories, roads and towns. Very often, the space and some of the materials that are needed for creating the infrastructure are obtained by deforestation.)

Answers to Review questions

1 **a** Iain had more pupils vote yes and Duncan had more pupils vote no.

b Charts will vary but the frequencies used should be Yes 27, No 22, Don't know 11.

2 **a** Yes, the sector for Indian men is twice the size of the sector for Indian women.

b No, $\frac{3}{8}$ of the visitors are female and $\frac{5}{8}$ are male.

3 **a** 100

b Nearly half of the seeds produced 8–9 potatoes when grown in straw, so growing in straw was effective.

4 **a**

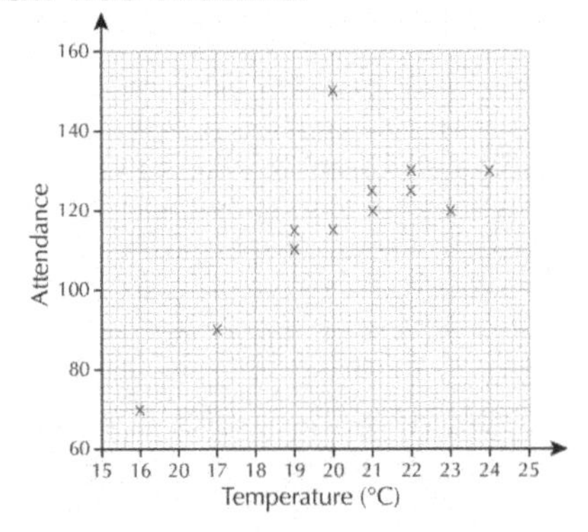

b positive correlation

c around 100

d (20, 150) for the last match – suitable reason such as the last match of the season

Answers to Challenge – Rainforest deforestation

1

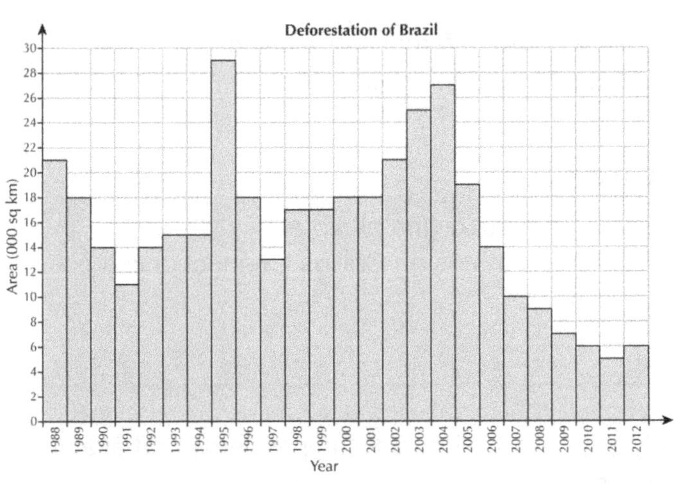

2 It was decreasing every year.

3 It was increasing every year.

4 **a** economic growth **b** economic slowdown

5 As the economy grows so does the rate of deforestation.

6 cattle ranches

7 20%

8

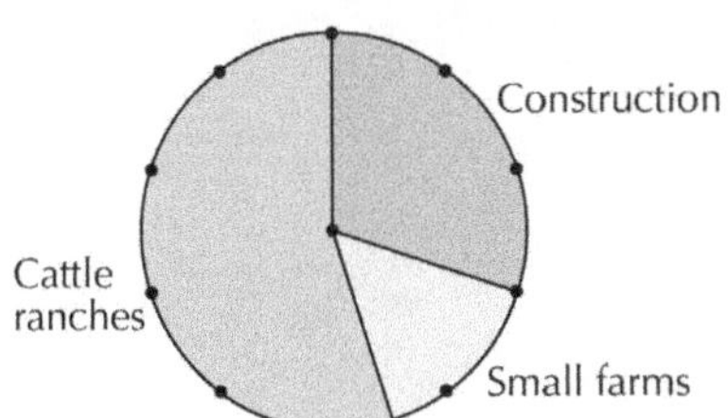

5 Circles

Learning objectives

- How to use π
- How to use π to calculate the circumference of a circle
- How to use π to calculate the area of a circle

Prior knowledge

- That the formula to work out the approximate length of the circumference of a circle is $C = 3d$

Context

- Tell pupils that the circle is probably the most important shape in the universe. It is also the most mysterious. We use a fascinating number that pupils may have heard of, called pi, written as π, which is used to calculate the circumference (perimeter) of a circle. But π cannot be written exactly as a number and its decimal places never end. Encourage pupils to prepare for this chapter by doing their own research on π. Encourage pupils to present their findings to the class.

Discussion points

- Ask pupils to name the radius, diameter and circumference of a circle.
- What is the relationship between the diameter and the radius of circle?

Associated Collins ICT resources

- Chapter 5 interactive activities on Collins Connect online platform
- *Clock*, *Stadium* and *Trapeze artists* Wonder of Maths on Collins Connect online platform

Curriculum references

Solve problems

- Develop their mathematical knowledge, in part through solving problems and evaluating the outcomes, including multi-step problems

Geometry and measure

- Calculate and solve problems involving: perimeters of 2D shapes (including circles), areas of circles and composite shapes

Fast-track for classes following a 2-year scheme of work

- Pupils will have met the formulae for area and circumference of a circle in Year 8. Check pupils' understanding by giving them some examples and go through the more formal explanation for area at the beginning of Lesson 5.2. If pupils are confident and fluent, move directly to Lesson 5.3.

Lesson 5.1 The formula for the circumference of a circle

Learning objective
- To calculate the circumference of a circle

Resources and homework
- Pupil Book 3.1, pages 83–86
- Online homework 5.1, questions 1–10

Links to other subjects
- **Design and technology** – to use the parts of a circle to describe the characteristic of a design involving circles
- **Physical education** – to use the language of circles to describe the features of a track or the patterns in a dance routine

Key word
- π

Problem solving and reasoning help
- **MR** question 5 and **PS** question 6 in Exercise 5A of the Pupil Book, plus the problem solving activity at the end of the exercise, give pupils the opportunity to use the language and characteristics of a circle that they have already learned. Pupils also have the opportunity to make links between the circumference of a circle and perimeter in other shapes. Pupils working at this level often fail to make these links.

Common misconceptions and remediation
- Pupils often confuse radius and diameter. Give them plenty of opportunity to use both. Collective memory activities would be useful in helping pupils to remember which is which.

Probing questions
- A circle has a diameter of 10 m. What is its circumference?
- A circle has a circumference of 140 cm. What is its diameter and its radius?

Part 1
- Draw a circle on the board, clearly marking the centre. Ask pupils to write a short definition of a circle. Share pupils' examples, then compare it to this example: *A two-dimensional shape made by drawing a curve that is always the same distance from a centre.*

Part 2
- Remind pupils of the formula for the circumference and area of a circle, as learned in Year 8.
- Then work through the alternative justification on page 83 of the Pupil Book. Take your time and encourage pupils to ask questions about anything they may be unsure of.
- **Pupils can now do Exercise 5A from Pupil Book 3.1.**

Part 3
- Working in pairs, ask pupils to write an example of a question involving the circumference of a circle, which has to do with real life.
- Encourage **more able** pupils to write multi-step questions that progress in difficulty.

- Discuss with the class what makes the questions easy or difficult. Try to draw out the idea of using inverse operations to add to the complexity of questions. You could generalise this to other areas of mathematics.

Answers

Exercise 5A

1 **a** 9.4 cm   **b** 17.3 cm   **c** 21.4 cm
d 6.3 m   **e** 11.0 m   **f** 13.5 m

2 **a** 18.8 cm   **b** 22.0 cm   **c** 45.2 cm
d 31.4 m   **e** 40.8 m   **f** 54.7 m

3

Coin	Diameter (mm)	Circumference
1p	20.3	64 mm
2p	25.9	81 mm
5p	18	57 mm
10p	24.5	77 mm
£1	22.5	71 mm
£2	28.4	89 mm

4 200 m

5 The square has the greatest perimeter. It is $4 \times 5 = 20$ cm while the circle is $6 \times \pi = 18.8$ cm.

6 239 m

Problem solving: To calculate the perimeter of a semicircle

A 7.7 cm

B 25.7 cm

Lesson 5.2 The formula for the area of a circle

Learning objective

- To calculate the area of a circle

Links to other subjects

- **Design and technology** – to use the definition of the circumference of a circle to describe the characteristic of a design involving circles
- **Physical education** – to use the language of circles to describe the features of a track
- **Science** – to use the definition of circumference to approximate the solution of distance–time problems involving circles in real life

Resources and homework

- Pupil Book 3.1, pages 86–90
- Online homework 5.2, questions 1–10

Key words

- No new key words for this topic

Problem solving and reasoning help

- The problem solving activity at the end of Exercise 5B in the Pupil Book gives pupils the opportunity to extend their understanding of circles to help them solve problems involving semicircles, which is something that pupils often struggle with.

Common misconceptions and remediation

- Pupils often do not make the link between the work they have done previously on perimeter and area, and the work on the circumference and area of a circle.

Probing questions

- What is the least information needed to be able to find the circumference and area of a circle?
- How would you go about finding the area of a circle if you know the circumference?

Part 1

- Start by revising the perimeters and areas, using familiar shapes such as rectangles.
- Make sure that each pair of pupils has a mini whiteboard or a sheet of A4 paper.
- Ask pairs to draw a rectangle with: perimeter 10 cm and area 6 cm^2, then to work out the area, and hold it up for all to see. The first pair to show the correct answer gains a point.
- Repeat the activity using different examples to suit the ability of the class.

Part 2

- Remind pupils of the formula for the circumference of a circle, which they learnt in Year 8. Introduce π on the calculator.
- Then work through examples 3 and 4 on pages 87 and 88 of the Pupil Book.
- **Pupils can now do Exercise 5B from Pupil Book 3.1.**

Part 3

- Give pupils some work with mistakes. Ask them to identify and correct the mistakes.

Answers

Exercise 5B

1 **a** 12.6 cm^2 **b** 78.5 cm^2 **c** 162.9 cm^2
d 3.1 m^2 **e** 0.3 m^2 **f** 91.6 m^2

2 **a** 19.6 cm^2 **b** 38.5 cm^2 **c** 55.4 cm^2
d 28.3 m^2 **e** 95.0 m^2 **f** 113.1 m^2

3 254 mm^2

4 Jenny has used the circumference formula.
Area = $\pi \times r^2$
= $\pi \times 16$
= 16π cm^2

5 The areas are 28.27 cm^2 and 113.10 cm^2, so Jackson is wrong. The area of his circle is 4 times the area of Finlay's circle.

6 346 cm^2

7 31 cm^2

Problem solving: Area of a semicircle

A 14.1 cm^2

B 39.3 cm^2

Lesson 5.3 Mixed problems

Learning objective

- To solve problems involving the circumference and area of a circle

Links to other subjects

- **Design and technology** – to use the formula of the circumference of a circle to complete designs incorporating circles
- **Science** – to use the definition of circumference to approximate the solution of distance–time problems involving circles in real life

Resources and homework

- Pupil Book 3.1, pages 90–91
- Online homework 5.3, questions 1–10

Key words

- No new key words for this topic

Problem solving and reasoning help

- **PS** question 6 of Exercise 5C in the Pupil Book requires learners to make decisions about what mathematics to use in a more complex real-life problem.

Common misconceptions and remediation

- Pupils often confuse the definitions of the parts of circles. Use activities like that in Lesson 5.1 to help pupils overcome this. Pupils also confuse the formulae, particularly in terms of the use of radius and diameter. This is often because they cannot visualise what they are doing. Activities like those in the previous lesson will help pupils to overcome this. Even at this level, it is important for pupils to learn to make independent choices about the mathematics they need to use.

Probing questions

- A touring cycle has wheels of diameter 75 cm. How many rotations does each wheel make for every 10 km travelled?
- Give pupils some work with mistakes. Ask them to identify and correct the mistakes.

Part 1

- Ask pupils to tell you the relationship they found for the area and circumference of a circle.
- Using this starting point, ask pupils if they can write similar formulae for half a circle (a semicircle) or a quarter of a circle?

Part 2

- Check that pupils are confident with the formulae and the variations in Part 1.
- Pupils should be ready to do Exercise 5C. Remind them they will need to make choices about which formulae to use.
- If necessary, go through an example. Stress the decisions you make about the mathematics that you use. Question 4 of the exercise might be a suitable example to use.
- **Pupils can now do Exercise 5C from Pupil Book 3.1.**

Part 3

- Ask pupils to work in small groups on the following problem:
  Sarah and Jane each have a circular rug in their bedrooms.
  Sarah's rug has a diameter of 3 m. Jane's rug has a diameter of 1.5 m.
  Sarah says that the circumference of her rug is twice the circumference of Jane's.
  Is she correct? Explain your answer.
- Encourage pupils to focus on explaining their answer. Then identify one group or more to present their answer to the class. The rest of the class could give formative feedback based on a set of agreed criteria. For example: accuracy of solution, quality of explanation, use of diagrams and formulae to support their argument.

Answers

Exercise 5C

1 18.8 m
2 4.5 m²
3 40 200 km
4 2 bottles
5 **a** 11 310 cm² **b** 1.13 m²
6 400 m

Review questions (Pupil Book pages 92–93)

- The review questions will help to determine pupils' abilities with regard to the material within Chapter 5.
- These questions also draw on the maths covered in earlier chapters of the book to encourage pupils to make links between different topics.
- The answers are on the next page of this Teacher Pack.

Financial skills – Athletics stadium (Pupil Book pages 94–95)

- This activity is designed to give pupils the opportunity to apply their knowledge to a multi-step, real-life problem. The context is familiar, but the activity is presented in a slightly more complex way than pupils may be used to.
- All the information required to answer the questions is in the text, but pupils will need to read and then think carefully about how they access the information. Remind them to highlight the key information they will need. Tell pupils that they will need to combine their skills not only from this chapter, but also from different areas of mathematics, for example, number.
- Before the class starts working on questions 1 to 5 in the Pupil Book, ask them some questions as a warm-up. Here are some examples:
  - Can you think of everyday examples of when you might need to use the formulae for the circumference and/or area of a circle?
  - Can you think of any sports that use circles in the design of their tracks and/or pitches?
- Pupils can now work on the questions individually or in groups.
- Encourage pupils to develop this topic further by using the internet to research other sports. Groups or individuals could present their findings to the class.

Answers to Review questions

1 **a i** 31.4 cm **ii** 78.5 cm²
b i 37.7 cm **ii** 113.1 cm²
c i 25.1 m **ii** 50.3 m²
2 **a** 12.6 cm **b** 12.6 cm²
Only the units are different.
3 1260 m
4 The circle – it has an area of 201 cm², while the area of the rectangle is 200 cm².
5 **a** 9.4 cm **b** 530
6 15.9 cm

Answers to Financial skills – Athletics stadium

1 **a** 6.3 m **b** 3.1 m² **c** £99
2 **a** 353 m² **b** £21 000
3 **a** 3200 m² **b** £11 500
4 **a** 12 m³ **b** £540
5 **a** 1.6 m³ **b** £80

6 Enlargements

Learning objectives

- How to use a scale factor to show an enlargement
- How to use rays to enlarge a shape about a centre of enlargement
- How to enlarge a shape about a centre of enlargement on a coordinate grid.

Prior knowledge

- How to plot coordinates
- How to work out the areas of 2D shapes

Context

- This chapter starts by showing pupils how to enlarge a 2D shape by a positive whole number scale factor. Pupils are then shown how to enlarge a shape using a centre of enlargement before being taught how to use a coordinate grid to enlarge a shape. Using photographs in the lesson may help their understanding.

Discussion points

- Why would we want to enlarge a shape or an object? When would we use this skill?
- Are you able to give me some examples?
- How could you find the scale factor and centre of enlargement from a completed enlargement?

Associated Collins ICT resources

- Chapter 6 interactive activities on Collins Connect online platform
- *Using enlargement* and *Using proportion and scale factors* videos on Collins Connect online platform

Curriculum references

Reason mathematically

- Extend and formalise their knowledge of ratio and proportion in working with measures and geometry, and in formulating proportional relations algebraically

Ratio, proportion and rates of change

- Use scale factors, scale diagrams and maps

Geometry and measures

- Use the standard conventions for labelling the sides and angles of triangle ABC
- Identify and construct congruent triangles, and construct similar shapes by enlargement, with and without coordinate grids

Fast-track for classes following a 2-year scheme of work

- If pupils in the class grasp concepts quickly, then it will be possible for you to combine Lesson 6.1 and Lesson 6.2. Encourage **more able** pupils to move straight to the more challenging questions towards and at the end of each exercise in this chapter.

Lesson 6.1 Scale factors and enlargements

Learning objective

- To use a scale factor to show an enlargement

Links to other subjects

- **Art** – to enlarge a drawing
- **Design and technology** – to enlarge a design by a scale factor

Resources and homework

- Pupil Book 3.1, pages 97–101
- Online homework 6.1, questions 1–10

Key words

- enlarge
- scale factor
- enlargement
- similar

Problem solving and reasoning help

- **MR** question 5 of Exercise 6A in the Pupil Book requires pupils to look at a shape and then determine which one of three other shapes is the correct enlargement. Remind pupils that all sides need to be enlarged by the given scale factor, in this case: × 3. **PS** question 6 relates to the enlargement of a car and the calculation of the scale factor. **Less able** pupils may struggle with the dimensions, as they are different units.

Common misconceptions and remediation

- Pupils sometimes do not enlarge all the lines, or they may enlarge by an incorrect scale factor. Remind pupils that the shape that is produced after making an enlargement will look similar to the original; it will just be a bigger or smaller version.

Probing questions

- What changes when you enlarge a shape?
- What stays the same when you enlarge a shape?

Part 1

- Imagine a cube that has an edge length of 2 cm. What is the volume of the cube? (8 cm^3)
- Now imagine the cube is twice as big. What is its volume now? (64 cm^3)
- Make sure pupils grasp that 'twice as big' means: the edges of the cube are multiplied by 2.
- Extend this activity by asking pupils to make the original cube three or four times larger. Allow pupils to work out their answers on paper.

Part 2

- Show pupils the following shapes, or shapes that are similar.

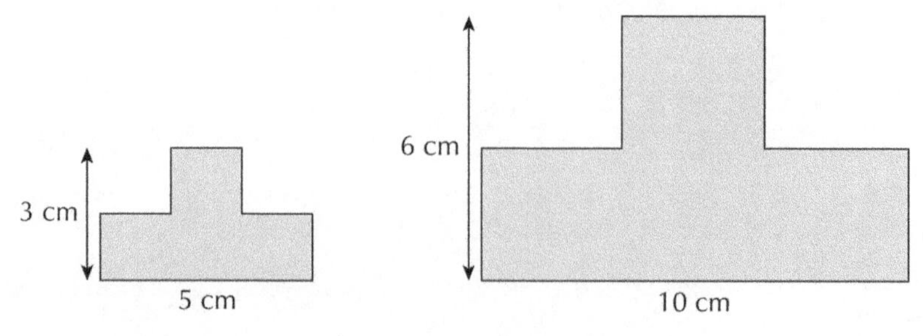

- Explain to pupils that shape Y is an enlargement with a 'scale factor of 2 of shape X'. Ask pupils to explain how they know that this is correct.

- Then draw the following shape on the board and ask pupils what the length of each side would be if it was enlarged by a scale factor of 3.

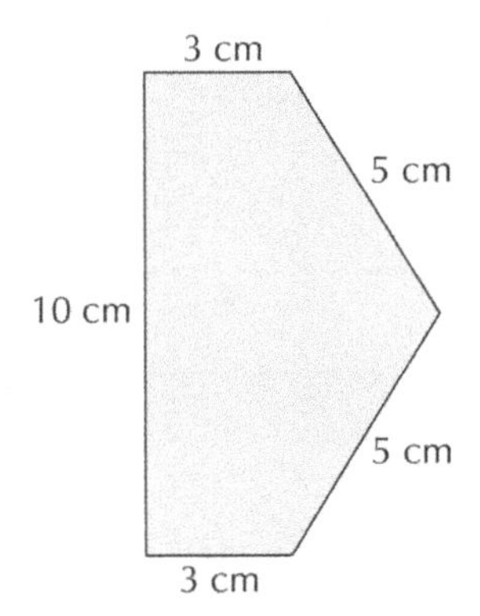

- Ask pupils if the angles inside the shape will also be enlarged. (No.) Then ask why.
- **Pupils can do Exercise 6A from Pupil Book 3.1.**

Part 3

- Ask pupils to describe which properties of a shape change after an enlargement and which properties remain the same, for example: length of sides change; area changes; all angles stay the same size.

Answers

Exercise 6A

1 **a** 3 **b** 2 **c** 4 **d** 5

2 **b**, they are the only pair that are enlargements of each other

3 DE = 12 cm, EF = 9 cm, DF = 12 cm

4 EF = 10 cm, FG = 16 cm, GH = 22 cm, HE = 16 cm

5 **a** rectangle B

b rectangle C, 4 × 5 cm = 20 cm, but 4 × 9 cm = 36 cm, not 32 cm

6 10

Challenge: Algebra with enlargements

A $x = 8$ cm

B $x = 5$ cm

Lesson 6.2 The centre of enlargement

Learning objective

- To enlarge a shape about a centre of enlargement

Resources and homework

- Pupil Book 3.1, pages 102–105

Links to other subjects

- **Art** – to enlarge a drawing using a centre of enlargement
- **Design and technology** – to enlarge a design by a scale factor

Key words

- centre of enlargement
- ray

Problem solving and reasoning help

- **PS** question 5 of Exercise 6B in the Pupil Book requires pupils to enlarge a shape that has the centre of enlargement inside the shape. Pupils find this type of question difficult, as they are unsure where to draw the lines. Remind pupils to follow the method they have been taught.

Common misconceptions and remediation

- Pupils often use an incorrect point as the centre of enlargement. Or, pupils simply enlarge the shape without reference to the given point.

Probing questions

- What information do you need to be given in order to enlarge a shape?
- How can you check that a shape has been correctly enlarged from a centre of enlargement?

Part 1

- On the board, draw a grid similar to the one shown here.
- Ask individual pupils to complete any cell in the multiplication grid. Ask these pupils if they have any particular strategies for working out their answers.
- Discuss some of the strategies used. For example, leave out the decimal point, then multiply and the answer will have one decimal place; to multiply by 4, double the number and then double again.

	× 2	**× 3**	**× 4**
1.2			
2.5			
3.4			
5.6			
7.9			

Part 2

- Remind the class that the three transformations they have met so far (reflections, rotations and translations) do not change the size of an object. They are now going to look at a transformation that does change the size of an object: an *enlargement*.
- Draw the diagram, as shown, on the board.
- Explain to pupils that each side of ΔA′B′C′ is twice as long as the corresponding side of ΔABC and that OA′ = 2 × OA, OB′ = 2 × OB and OC′ = 2 × OC. ΔABC has been enlarged by a *scale factor* of two about the *centre of enlargement*, O, to give the *image* A′B′C′. The

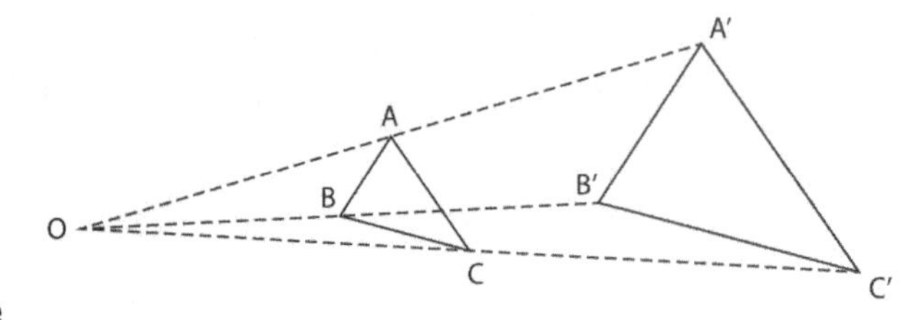

dashed lines are called the *guidelines* or *rays* for the enlargement. Emphasise that to enlarge a shape, a centre of enlargement and a scale factor are needed.

- Next, show the class how to enlarge the triangle XYZ (shown on the right) by a scale factor of two about the centre of enlargement, O. Draw rays OX, OY and OZ. Measure the length of each ray and multiply it by two. Extend each ray to its new length measured from O and plot the points X′, Y′ and Z′. Join X′, Y′ and Z′.
- ΔX′Y′Z′ is the enlargement of ΔXYZ by a scale factor of 2 about the centre of enlargement, O.
- **Pupils can now do Exercise 6B from Pupil Book 3.1.**

Part 3

- Show pupils two shapes, one an enlargement of the other, and ask them to work out where the centre of enlargement is by drawing lines.
- What is the minimum number of lines they will need to draw?

Answers

Exercise 6B

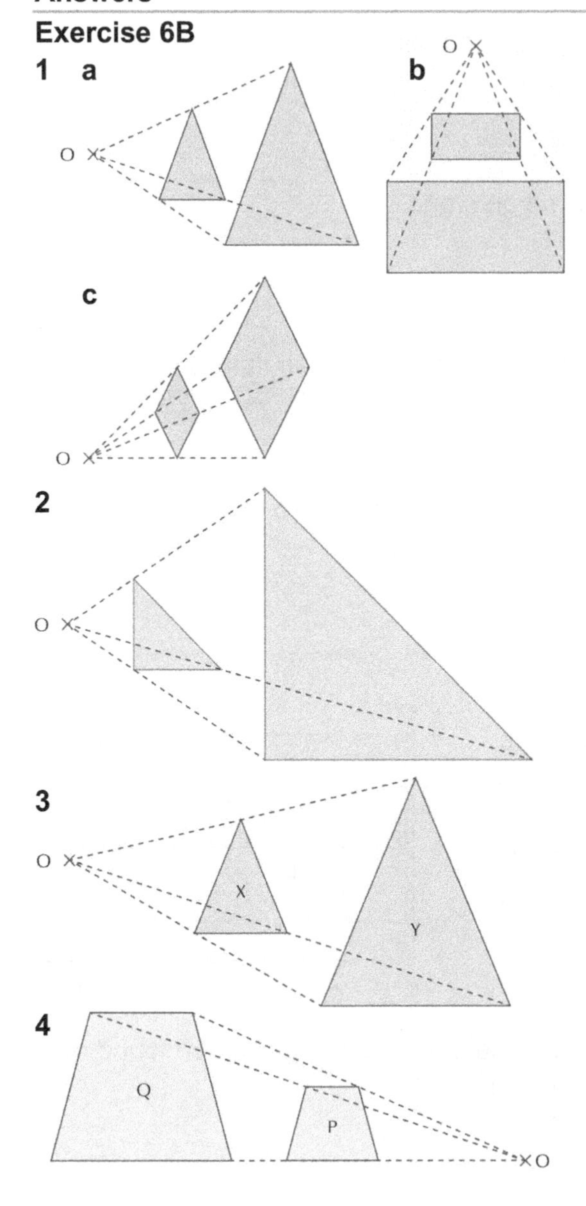

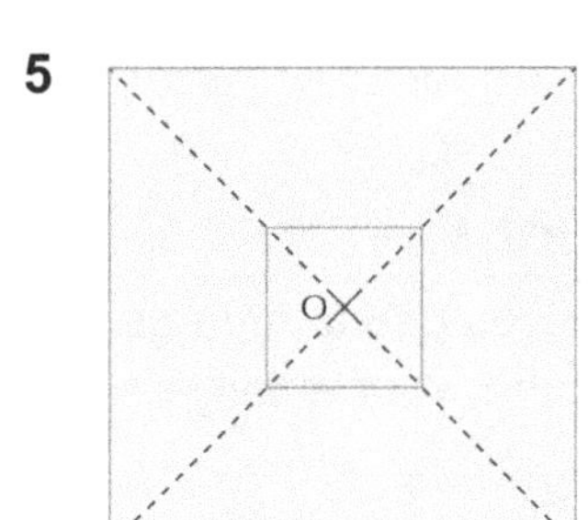

Reasoning: Reductions

A

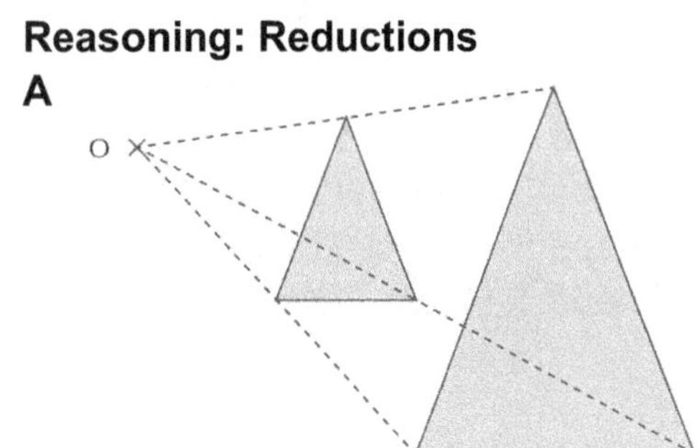

B

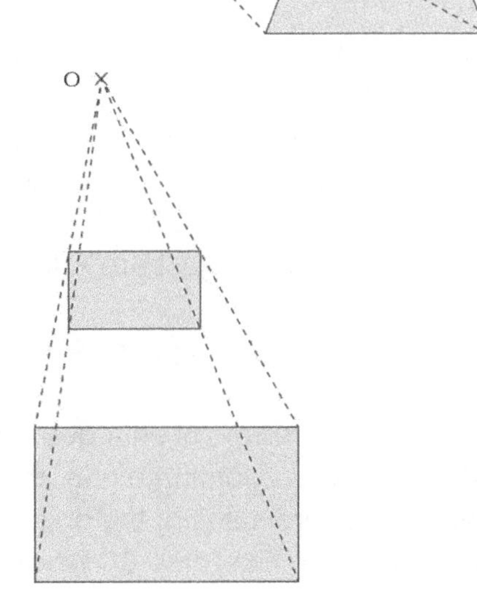

Lesson 6.3 Enlargements on grids

Learning objective

- To enlarge a shape on a coordinate grid

Links to other subjects

- **Art** – to enlarge a drawing using a grid

Resources and homework

- Pupil Book 3.1, pages 106–109
- Online homework 6.3, questions 1–10

Key words

- No new key words for this topic

Problem solving and reasoning help

- **MR** question 6 of Exercise 6C in the Pupil Book requires pupils to draw axes and construct the shape to be enlarged from the given coordinate points. Pupils sometimes struggle to do this accurately. Put pupils in pairs so that they can check each other's axes and shapes and make sure that each pupil has drawn them correctly. **Less able** pupils will need help with drawing the correct axes.

Common misconceptions and remediation

- Pupils will occasionally label axes incorrectly and plot points the wrong way around. Reinforce the correct methods for drawing axes and plotting points using Part 1 of this lesson.

Probing questions

- What method would you use to enlarge a diagonal line on a grid?
- How can you check that your enlargement is correct?

Part 1

- Ask pupils to draw a 20 by 20 grid on squared paper. Plot each point. Join them with straight lines and then name the resulting shape.
  - Shape 1: (9,16), (15,16), (18,20), (12,20)
  - Shape 2: (13,6), (13,2), (19,2)
  - Shape 3: (11,14), (13,14), (15,12), (15,10), (13,8), (11,8), (9,10), (9,12)
  - Shape 4: (4,19), (1,16), (2,13), (6,13), (7,16)

Part 2

- Show the class how to enlarge a shape about the origin on a coordinate grid.
- The *object* rectangle ABCD on the coordinate grid shown is enlarged by a scale factor of three about the origin O, to give the *image* rectangle A′B′C′D′.
- The coordinates of the object are A(0, 2), B(3, 2), C(3, 1) and D(0, 1).
- The coordinates of the image are A′(0, 6), B′(9, 6), C′(9, 3) and D′(0, 3).

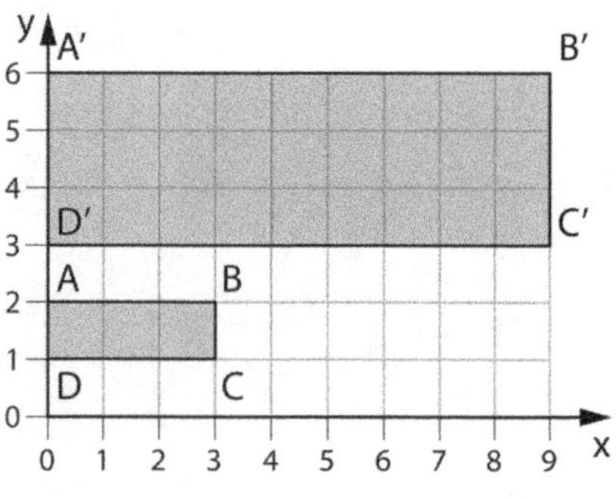

- Notice that when a shape is enlarged by a scale factor about the origin of a coordinate grid, the coordinates of the enlarged shape can be found by multiplying the coordinates of the object shape by the scale factor.
- **Pupils can now do Exercise 6C from Pupil Book 3.1.**

Part 3

- Ask pupils the following question, which leads into the problem solving activity on photographs at the end of this chapter.
- Kian has taken a photograph of a lion and wants to enlarge it so that it will fit into a frame that is 26 cm by 18 cm. The photo measures 6.5 cm high by 4.5 cm wide.
- By what scale factor should Kian multiply the photo?

Answers

Exercise 6C

1 **a**

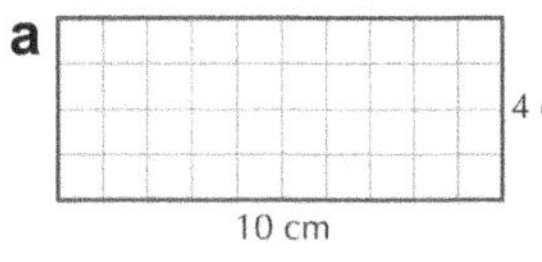

b 6 cm, 8 cm

c

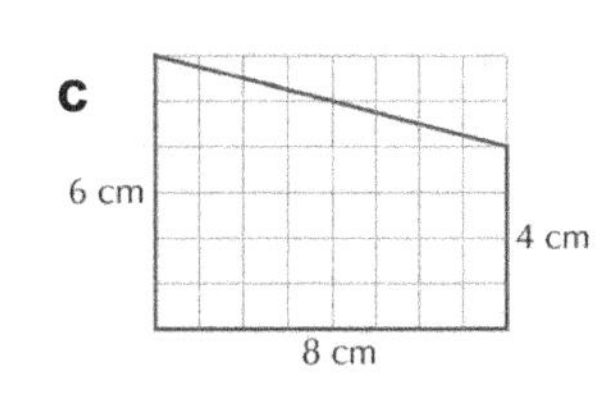

d 8 cm, 2 cm, 6 cm

2 **a**

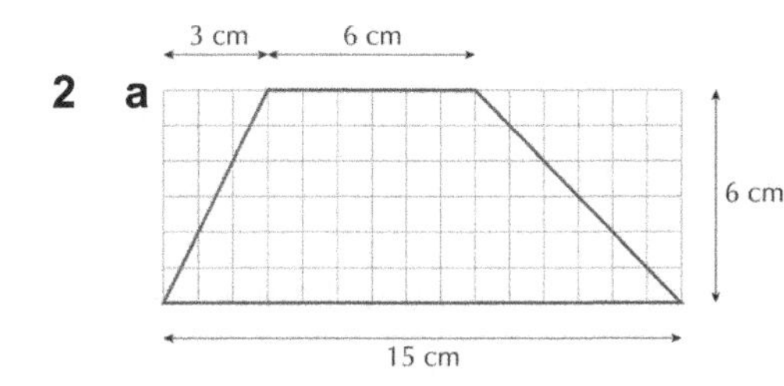

b

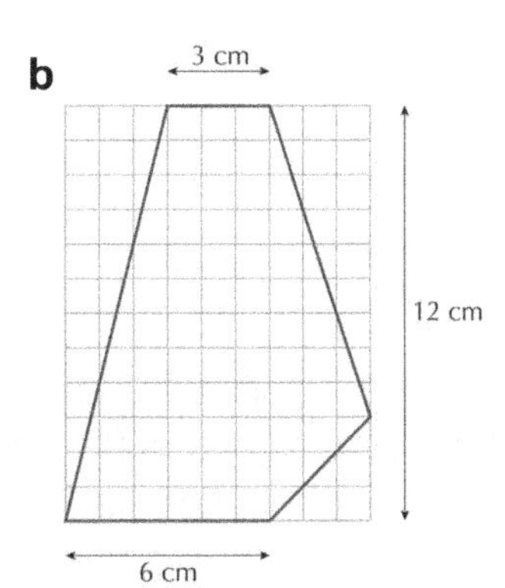

3 **a** vertices at (8, 6), (8, 2), (4, 2)
b DE = 12 cm, EF = 9 cm, DF = 12 cm
c vertices at (3, 9), (6, 9), (6, 6), (9, 6), (9, 9), (12, 9), (12, 3), (3, 3)
d vertices at (0, 8), (8, 8), (8, 12), (12, 6), (8, 0), (8, 4), (0, 4)

4 vertices at (6, 12), (12, 12), (9, 3), (3, 3)

5 **a** 2 **b** (9, 1)

6 **a**

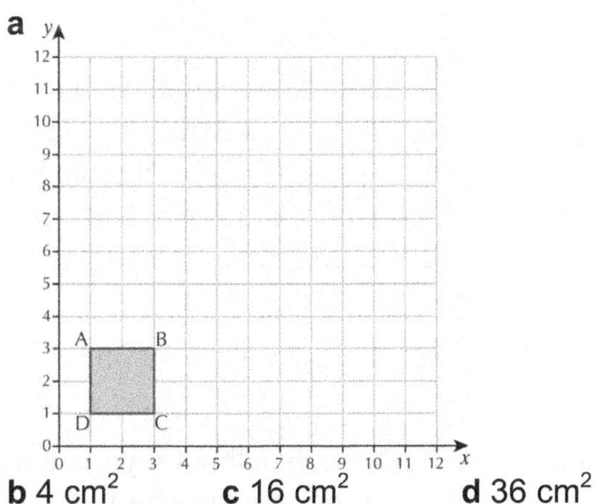

b 4 cm^2 **c** 16 cm^2 **d** 36 cm^2 **e** 64 cm^2
f the area scale factor is the square of the scale factor

Activity: Enlarged stickmen

Check pupils' posters.

Review questions (Pupil Book pages 110–111)

- The review questions will help to determine pupils' abilities with regard to the material within Chapter 6.
- These questions also draw on the mathematics covered in earlier chapters of the book to encourage pupils to make links between different topics.
- The answers are on the next page of this Teacher Pack.

Problem solving – Photographs (Pupil Book pages 112–113)

This activity consolidates topics previously covered on extracting data, area and ratio.

- Draw the squares and rectangles on the board as shown.
- Ask pupils to answer these questions:
    - Which two shapes have the same perimeter? (E and G)
    - Which two shapes have the same area? (D and F)
    - Which two shapes have the same value for their perimeter and area? (B and C)

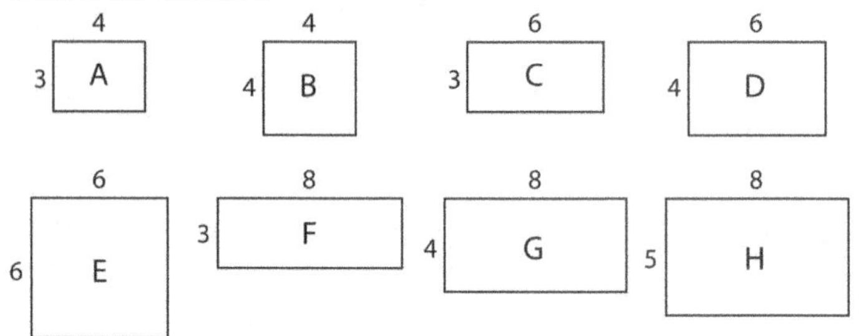

- Check pupils' answers after each question to ensure that they understand the difference between perimeter and area.
- Encourage pupils to suggest questions that might arise from the data in the activity, for example, the use of inches for the dimensions of the prints. Discuss the imperial units that are still being used. Examples are weighing vegetables at market stalls (pounds and ounces) and human weight (stone).
- Discuss the reasons why shops encourage us to buy in bulk.
- It is important for pupils to realise that when we refer to 'photograph enlargements', they may not actually be true mathematical enlargements. It may be useful to revise enlargements and scale factors before pupils start question 3 of the activity.
- Discuss 'best buys' with pupils. Do people really compare prices in different shops? Ask pupils for their views. This may well lead to some discussion about the influence of advertisements on television and on the internet.
- Before starting question 5 on golden rectangles, it may be useful to revise simplifying ratios.
- Golden rectangles appear in paintings and buildings, both ancient and modern. For example, the painting by Salvador Dali, *The Sacrament of the Last Supper* is a golden rectangle and the exterior of the Parthenon on the Acropolis in Athens is based on the golden rectangle.
- Direct pupils to the internet where there is a wealth of information on golden rectangles and the golden ratio, Φ.
- $\Phi = \frac{\sqrt{5}+1}{2} = 1.618\ 039\ 887\ 498\ 948\ 482\ 045\ 868\ 343\ 656\ 381\ 177\ 203\ 091\ 798\ 057\ 6\ \ldots$
- Let pupils work through questions 1 to 5 in the activity.
- You could end by asking the class for the different areas of mathematics that they have used during the lesson. Individual pupils could write these on the board or on a flip-chart.

Answers to Review questions

1 **a** 2 **b** 4

2 C is the odd one out. A, B and D are all enlargements/similar.

3 **a**

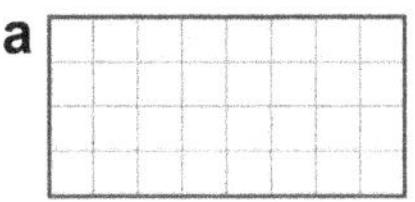

b

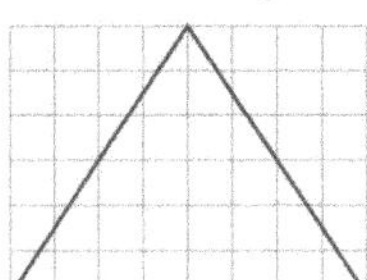

4

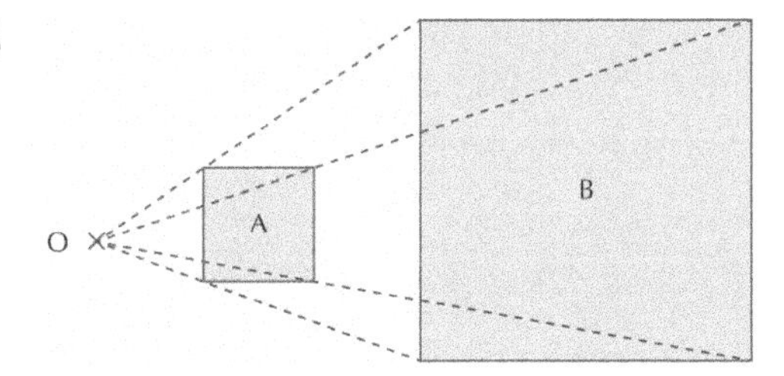

5

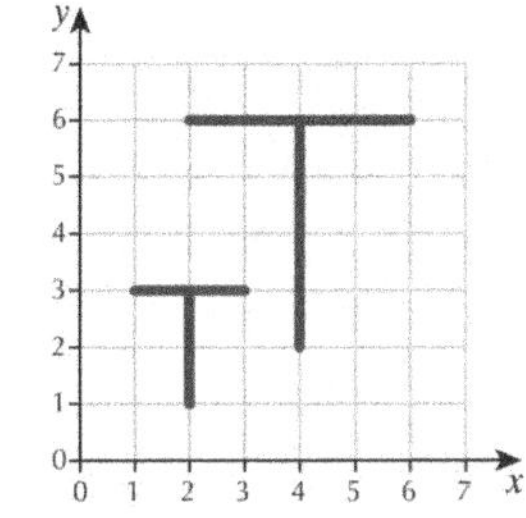

6 **a** A(2, 4), B(4, 4), C(4, 2), D(2, 2 **b** Check pupils' drawings.
c A'(4, 8), B'(8, 8), C'(8, 4), D'(4, 4)
d A"(6, 12), B"(12, 12), C"(12, 6), D"(6, 6)

Answers to Problem solving – Photographs

1 £191.25

2 **a** 13 square inches **b** 11 square inches **c** 22 square inches

3 12" × 8" and 6" × 4", with a scale factor of 2

4 **a** 6" × 4" and 7" × 5" **b** Fast Print, £4.50

5 **a**

6" × 4"	6 : 4	6 ÷ 4 = 1.5	1.5 : 1
7" × 5"	7 : 5	7 ÷ 5 = 1.4	1.4 : 1
8" × 6"	8 : 6	8 ÷ 6 = 1.33	1.33 : 1
12" × 8"	12 : 8	12 ÷ 8 = 1.5	1.5 : 1
13" × 8"	13 : 8	13 ÷ 8 = 1.625	1.625 : 1

b the 13" × 8" print

7 Fractions

Learning objectives

- How to subtract any two fractions
- How to multiply any two fractions
- How to divide any two fractions

Prior knowledge

- How to multiply or divide a fraction by an integer

Context

- The number system was originally used simply for counting, and only positive whole numbers were used. Later, the number system was extended to include zero. Zero is very important as a place holder, and in negative numbers and fractions. By now, pupils should have an understanding of the ordinal value of fractions as well as their use as operators. This chapter builds on the Year 8 work on fractions, leading pupils to using fractions to solve real-life problems.

Discussion points

- Which sets of equivalent fractions do you know?
- Give me a fraction between $\frac{1}{5}$ and $\frac{1}{2}$. How did you find it? To which fraction is it closer? How do you know this?

Associated Collins ICT resources

- Chapter 7 interactive activities on Collins Connect online platform
- *Dividing fractions* and *Ratio* video on Collins Connect online platform
- *Golden Ratio* Wonder of Maths on Collins Connect online platform

Curriculum references

Solve problems

- Develop their mathematical knowledge, in part through solving problems and evaluating the outcomes, including multi-step problems
- Develop their use of formal mathematical knowledge to interpret and solve problems, including in financial mathematics

Number

- Use the four operations, including formal written methods, applied to proper and improper fractions and mixed numbers

Fast-track for classes following a 2-year scheme of work

- The material in Lesson 7.1 should be familiar to pupils but pupils working at this level are likely to need reinforcing of the work. Check pupils' understanding by working through some examples. Only if appropriate, move on to Lesson 7.2.

Lesson 7.1 Adding and subtracting fractions

Learning objective

- To add or subtract any two fractions

Resources and homework

- Pupil Book 3.1, pages 115–118
- Homework Book 3, section 7.1
- Online homework 7.1, questions 1–10

Links to other subjects

- **Food technology** – to calculate dietary content
- **Art** – to calculate the layout of a painting, for example, using the rule of thirds

Key words

- No new key words for this topic

Problem solving and reasoning help

- **PS** questions 7 and 8 in Exercise 7A of the Pupil Book require pupils to solve more complex problems. **MR** question 10 requires pupils to explain and justify their solutions. You could do this question as a class using stems such as 'I agree with … or disagree with … because …' The challenge at the end of the exercise gives pupils the opportunity to make links across other strands of mathematics, in this case sequences.

Common misconceptions and remediation

- Pupils often get taught rules without fully understanding them, so some pupils may struggle to apply the rules in different contexts. These pupils often confuse the rules for adding and subtracting fractions with those for multiplying and dividing fractions.
- Another common problem is that pupils add or subtract the denominators as well as the numerators when adding and subtracting fractions.

Probing questions

- Why are equivalent fractions important when adding or subtracting fractions?
- What strategies do you use to find a common denominator when adding or subtracting fractions?
- Is there only one possible common denominator? What happens if you use a different common denominator?

Part 1

- Use a target board, as shown. Start by asking groups or individuals to give the first five multiples of various numbers.
- Once the idea of the multiple is well established, ask for the *lowest common multiple (LCM)* of a pair of numbers. (Recap the definition if necessary.) Continue for as long as necessary.

2	5	7	8
24	15	18	3
6	9	27	14
12	20	10	25

Part 2

- This is essentially a lesson on adding and subtracting fractions with different denominators.
- Pupils may have met this before, so you will probably just need to remind them of the basic rules and outline some more of the advanced techniques.
- Ask pupils how to work out $\frac{5}{6} + \frac{3}{4}$. Recall the method of finding the LCM (12) and converting the original fractions to equivalent fractions.

- Make sure pupils know how to deal with negative fractions, using **b** of Example 1 on page 115 of the Pupil Book. If needed, provide another example for pupils to try in pairs. Check their outcomes before moving on.
- **Pupils can now do Exercise 7A from Pupil Book 3.1.**

Part 3

- Give pupils some examples of addition and subtraction of fractions with common mistakes. Ask pupils to talk you through the mistakes and how they would correct them.

Answers

Exercise 7A

1 **a** $\frac{9}{12}$ **b** $\frac{2}{10}$ **c** $\frac{6}{9}$ **d** $\frac{5}{10}$

e $\frac{36}{40}$ **f** $\frac{24}{28}$ **g** $\frac{4}{32}$ **h** $\frac{4}{32}$

2 **a** $\frac{2}{7}$ **b** $\frac{2}{9}$ **c** $\frac{3}{5}$ **d** $\frac{2}{3}$ **e** $\frac{6}{10} = \frac{2}{3}$

f $\frac{4}{8} = \frac{1}{2}$ **g** $\frac{6}{9} = \frac{2}{3}$ **h** $\frac{6}{12} = \frac{1}{2}$ **i** $\frac{6}{8} = \frac{3}{4}$

3 **a** $\frac{4}{7}$ **b** $\frac{4}{9}$ **c** $\frac{2}{5}$ **d** $\frac{1}{3}$ **e** $\frac{4}{10} = \frac{2}{5}$

f $\frac{2}{8} = \frac{1}{4}$ **g** $\frac{3}{9} = \frac{1}{3}$ **h** $\frac{6}{12} = \frac{1}{2}$ **i** $\frac{10}{15} = \frac{2}{3}$

4 **a** $\frac{3}{10}$ **b** $\frac{3}{6} + \frac{1}{6} = \frac{4}{6} = \frac{2}{3}$ **c** $\frac{1}{8} + \frac{4}{8} = \frac{5}{8}$

d $\frac{2}{4} + \frac{1}{4} = \frac{3}{4}$ **e** $\frac{6}{8} + \frac{1}{8} = \frac{7}{8}$ **f** $\frac{6}{10} + \frac{1}{10} = \frac{7}{10}$

g $\frac{5}{12} + \frac{2}{12} = \frac{7}{12}$ **h** $\frac{2}{14} + \frac{1}{14} = \frac{3}{14}$ **i** $\frac{9}{15} + \frac{4}{15} = \frac{13}{15}$

5 **a** $\frac{9}{10} - \frac{2}{10} = \frac{7}{10}$ **b** $\frac{2}{6} - \frac{1}{6} = \frac{1}{6}$ **c** $\frac{4}{8} - \frac{1}{8} = \frac{3}{8}$

d $\frac{2}{4} - \frac{1}{4} = \frac{1}{4}$ **e** $\frac{6}{8} - \frac{1}{8} = \frac{5}{8}$ **f** $\frac{8}{10} - \frac{1}{10} = \frac{7}{10}$

g $\frac{11}{12} - \frac{10}{12} = \frac{1}{12}$ **h** $\frac{2}{14} - \frac{1}{14} = \frac{1}{14}$ **i** $\frac{12}{15} - \frac{2}{15} = \frac{10}{15} = \frac{2}{3}$

6 **a** $\frac{9}{20}$ **b** $\frac{5}{8}$ **c** $\frac{19}{20}$ **d** $\frac{7}{18}$

e $\frac{1}{20}$ **f** $\frac{7}{24}$ **g** $\frac{11}{20}$ **h** $\frac{1}{6}$

7 **a** $\frac{2}{3}$ **b** 100

8 **a** $\frac{1}{6}$ **b** 200

9 **a** $\frac{7}{8}$ **b** $\frac{3}{8}$

10 The perimeter is $\frac{2}{3} + \frac{2}{8} = \frac{11}{12}$ m, so Eve is correct.

11 **a** $\frac{31}{40}$ **b** $\frac{1}{40}$ **c** $\frac{7}{20}$ **d** $\frac{-17}{40}$

Challenge: Interesting fractions

A **a** $\frac{3}{4}$ **b** $\frac{3}{8}$ **c** $\frac{3}{16}$

B $\frac{3}{32}$

C **a** $\frac{4}{9}$ **b** $\frac{4}{27}$ **c** $\frac{4}{81}$

D $\frac{1}{81} + \frac{1}{243} = \frac{4}{243}$

Lesson 7.2 Multiplying fractions

Learning objective

- To multiply two fractions

Resources and homework

- Pupil Book 3.1, pages 118–121
- Homework Book 3, section 7.2
- Online homework 7.2, questions 1–10

Links to other subjects

- **Food technology** – to calculate dietary content
- **Geography** – to work out distances or areas from a map

Key words

- No new key words for this topic

Problem solving and reasoning help

- **PS** questions 4 and 5 in Exercise 7B of the Pupil Book require pupils to solve more complex problems. **MR** questions 7 and 8 require pupils to explain and justify their solutions. You could do one of the questions as a class using stems such as: 'I agree with … disagree with … because …'. The investigation at the end of the exercise gives pupils the opportunity to look at patterns and use inverse operations.

Common misconceptions and remediation

- Pupils often get taught rules without fully understanding them. Pupils may struggle to apply the rules in different contexts and often confuse the rules for multiplying and dividing fractions with those for adding and subtracting fractions.

Probing questions

- 'Multiplying makes numbers bigger.' When is this true and when is it false?

Part 1

- Use a target board such as the one shown. Choose individuals and ask each pupil to cancel a fraction to its simplest terms.

$\frac{5}{10}$	$\frac{8}{10}$	$\frac{8}{16}$	$\frac{4}{16}$	$\frac{4}{12}$
$\frac{6}{12}$	$\frac{6}{10}$	$\frac{24}{10}$	$\frac{24}{12}$	$\frac{14}{12}$
$\frac{14}{21}$	$\frac{7}{21}$	$\frac{7}{14}$	$\frac{12}{14}$	$\frac{12}{15}$
$\frac{5}{15}$	$\frac{5}{25}$	$\frac{20}{25}$	$\frac{20}{16}$	$\frac{14}{16}$

Part 2

- This is the first time that pupils will have met multiplication of a fraction by another fraction.
- One way to introduce this topic is to use calculators to investigate the answers to products such as $\frac{1}{5} \times \frac{3}{5}$, $\frac{5}{7} \times \frac{1}{4}$ and $\frac{3}{4} \times \frac{3}{8}$. Pupils will see the rule very quickly.
- Visual images using familiar examples such as $\frac{1}{2} \times \frac{1}{4}$ are also useful.
- Now repeat with $\frac{1}{2} \times \frac{4}{5}$, $\frac{3}{4} \times \frac{8}{9}$ and $\frac{4}{15} \times \frac{3}{8}$. Ask why the rule does not appear to work.
- Discuss cancelling in the initial product and in the answer.

- Explain that it is better to cancel the initial fractions, as this makes the calculations easier and means that the answer does not need to be cancelled down. Demonstrate with:
  $\frac{\cancel{4}^{1}}{\cancel{9}^{3}} \times \frac{\cancel{25}^{5}}{\cancel{28}^{7}} \times \frac{\cancel{3}^{1}}{\cancel{10}^{2}} = \frac{5}{42}$
- **Pupils can now do Exercise 7B from Pupil Book 3.1.**

Part 3

- Say: 'In the real world, mathematics problems are often surrounded by words. Part of the skill of solving a problem is finding the words to identify the mathematics required'.
- To consolidate and apply the learning from this lesson, ask pupils to work in pairs on this problem: 'At Sarah's house $\frac{3}{5}$ of the garden is grass. Sarah wants to use half of this to build a pond. What fraction of the garden is pond?'

Answers

Exercise 7B

1 **a** 9 **b** 23 **c** 30 **d** 7
e 11 **f** 7 **g** 9 **h** 7

2 $\frac{1}{3}$ of 30 → 10
$\frac{1}{4}$ of 24 → 6
$\frac{1}{5}$ of 35 → 7
$\frac{2}{3}$ of 27 → 18
$\frac{3}{4}$ of 36 → 27

3 **a** 4 **b** 8 **c** 10 **d** 30
e 2 **f** 10 **g** 2 **h** 14

4 £200

5 18

6 **a** $\frac{1}{12}$ **b** $\frac{1}{10}$ **c** $\frac{3}{16}$ **d** $\frac{4}{15}$
e $\frac{3}{16}$ **f** $\frac{3}{10}$ **g** $\frac{5}{12}$ **h** $\frac{21}{40}$

7 $\frac{3}{4} \times \frac{5}{8} = \frac{15}{32}$, which is bigger than $\frac{3}{8} \times \frac{4}{5} = \frac{12}{40}$.

8 Andrew is correct: $\frac{1}{2}$ of $\frac{1}{2} = \frac{1}{4}$, while $\frac{1}{4}$ of $\frac{1}{4} = \frac{1}{16}$.

Investigation: Multiplication of fractions

A **a** $\frac{2}{5}$ **b** $\frac{4}{5}$ **c** $\frac{3}{5}$

B **a** $\frac{12}{5}$ **b** $\frac{6}{7}$ **c** $\frac{4}{5}$

C **a** $\frac{16}{5}$ **b** $\frac{20}{7}$ **c** $\frac{8}{5}$

D **a** $\frac{32}{5}$ **b** $\frac{15}{7}$ **c** $\frac{6}{7}$

E **a** $\frac{1}{5}$ **b** $\frac{2}{5}$ **c** $\frac{6}{7}$

F **a** 1000 **b** 1000 **c** $\frac{10}{7}$

Lesson 7.3 Dividing fractions

Learning objective

- To divide one fraction by another

Resources and homework

- Pupil Book 3.1, pages 121–123
- Homework Book 3, section 7.4
- Online homework 7.3, questions 1–10

Links to other subjects

- **Food technology** – to calculate dietary content

Key word

- invert

Problem solving and reasoning help

- **PS** questions 7 and 8 in Exercise 7C of the Pupil Book require pupils to solve more complex problems. **MR** question 9 requires pupils to explain and justify their solutions. You could do this question as a class using stems such as: 'I agree with … disagree with … because …'. The challenge activity at the end of the exercise reinforces the use of inverse operations and starts to generalise using algebra.

Common misconceptions and remediation

- Pupils often get taught rules without fully understanding them. Pupils may struggle to apply the rules in different contexts and often confuse the rules for multiplying and dividing fractions with those for adding and subtracting fractions.

Probing questions

- How would you justify that $4 \div \frac{1}{5} = 20$? How would you use this to work out $4 \div \frac{2}{5}$?
- Do you expect the answer to be greater or less than 20? Why?

Part 1

- Working in pairs, ask pupils to consider this statement: 'Division makes things smaller'.
- When is this true and when is it false?
- Take feedback, and ask pupils to justify their responses.

Part 2

- Division is a little harder to see. If pupils use calculators, investigate problems such as: $\frac{2}{7} \div \frac{1}{3}$, $\frac{3}{4} \div \frac{4}{5}$ and $\frac{2}{3} \div \frac{8}{9}$. Some pupils may grasp the method.
- Pupils are unlikely to see that it is the same as turning the dividing fraction upside down and multiplying by it. Demonstrate this with the examples above. This method is easy to use but it is not an easy method to understand. The best way to explain the method is to use examples such as: How many halves are in 7? The answer is the same as multiplying by 2.
- Do more examples if necessary.
- **Pupils can now do Exercise 7C from Pupil Book 3.1.**

Part 3

- To summarise the learning in this chapter and consolidate previous learning, provide some examples of +, –, × and ÷ questions with common mistakes in them.
- Ask pupils to talk you through the mistakes and how they would correct them.

Exercise 7C

1 **a** $\frac{3}{8}$ **b** $\frac{4}{15}$ **c** $\frac{3}{40}$

d $\frac{3}{5} \times \frac{1}{4} = \frac{3}{20}$ **e** $\frac{5}{6} \times \frac{1}{2} = \frac{5}{12}$ **f** $\frac{3}{10} \times \frac{1}{4} = \frac{3}{40}$

g $\frac{4}{9} \times \frac{1}{9} = \frac{4}{27}$ **h** $\frac{4}{5} \times \frac{1}{7} = \frac{4}{35}$ **i** $\frac{7}{10} \times \frac{1}{5} = \frac{7}{50}$

2 **a** $\frac{15}{8}$ **b** $\frac{12}{10}$ **c** $\frac{18}{40}$

d $\frac{3}{5} \times \frac{3}{2} = \frac{9}{10}$ **e** $\frac{5}{6} \times \frac{5}{2} = \frac{25}{12}$ **f** $\frac{3}{10} \times \frac{8}{5} = \frac{24}{50}$

g $\frac{3}{8} \times \frac{5}{2} = \frac{15}{16}$ **h** $\frac{5}{8} \times \frac{3}{2} = \frac{15}{16}$ **i** $\frac{7}{10} \times \frac{9}{4} = \frac{63}{40}$

3 **a** $5 \times \frac{2}{1} = \frac{10}{1} = 10$ **b** $4 \times \frac{3}{1} = \frac{12}{1} = 12$ **c** $3 \times \frac{5}{1} = \frac{15}{1} = 15$

d $6 \times \frac{4}{1} = \frac{24}{1} = 24$ **e** $8 \times \frac{5}{1} = \frac{40}{1} = 40$ **f** $3 \times \frac{4}{1} = \frac{12}{1} = 12$

g $7 \times \frac{6}{1} = \frac{42}{1} = 42$ **h** $5 \times \frac{3}{1} = \frac{15}{1} = 15$

4 **a** $5 \times \frac{3}{2} = \frac{15}{2}$ **b** $7 \times \frac{3}{2} = \frac{21}{2}$ **c** $4 \times \frac{5}{3} = \frac{20}{3}$ **d** $8 \times \frac{5}{2} = \frac{40}{2}$

e $10 \times \frac{3}{2} = \frac{30}{2}$ **f** $4 \times \frac{4}{3} = \frac{16}{3}$ **g** $6 \times \frac{8}{5} = \frac{48}{5}$ **h** $9 \times \frac{8}{5} = \frac{72}{5}$

5 **a** $7 \times \frac{5}{3} = \frac{35}{3}$ **b** $4 \times \frac{5}{2} = \frac{20}{2}$ **c** $8 \times \frac{3}{2} = \frac{24}{2}$ **d** $9 \times \frac{5}{2} = \frac{45}{2}$

e $5 \times \frac{5}{2} = \frac{25}{2}$ **f** $5 \times \frac{8}{3} = \frac{40}{3}$ **g** $4 \times \frac{9}{5} = \frac{36}{5}$ **h** $7 \times \frac{8}{3} = \frac{56}{3}$

6 **a** ii has the smallest answer of $\frac{15}{4}$ ($3\frac{3}{4}$).

b i and iii both have the largest answer of $\frac{20}{3}$ ($6\frac{2}{3}$).

7 12 lengths

8 no, as $25 \div \frac{9}{10} = 27\frac{7}{9}$ (or $30 \times \frac{9}{10}$ cm = 27 cm)

9 $5 \div \frac{1}{3} = 5 \times \frac{3}{1} = 15 \neq \frac{3}{5}$, so James is correct.

Challenge: Algebra with fractions

A **a** 200 **b** 300 **c** 50

B **a** 300 **b** 800 **c** 12.5

Review questions

(Pupil Book pages 124–125)

- The review questions will help to determine pupils' abilities with regard to the material within Chapter 7.
- These questions also draw on the mathematics covered in earlier chapters of the book to encourage pupils to make links between different topics.
- The answers are on the next page of this Teacher Pack.

Problem solving – The 2016 Olympic Games in Rio

(Pupil Book pages 126–127)

- Pupils apply their understanding of fractions to a topical but more complex problem.
- Pupils need to work methodically to identify all the information required to answer the questions. Remind them to highlight the key information they will need.
- Pupils will also need to combine their understanding across fractions, decimals and percentages as well as their understanding of averages.
- Pupils should work in small groups to encourage discussions about the problem. Support groups by using guided group work if necessary.
- As a warm-up, you could ask some probing questions about fractions, such as:
  - What strategies do you use to find a common denominator when adding or subtracting fractions?
  - Is there only one possible common denominator? What happens if you use a different common denominator?
- Pupils could extend this activity by using the internet to explore other Olympic Games. For example: **http://www.olympic.org/olympic-results**, looks at Olympic records.
- Pupils could design their own question using fractions.

Answers to Review questions

1 **a** $\frac{4}{5}$ **b** $\frac{4}{10} = \frac{2}{5}$ **c** $\frac{4}{8} = \frac{1}{2}$

d $\frac{4}{8} = \frac{1}{2}$ **e** $\frac{4}{10} = \frac{2}{5}$ **f** $\frac{3}{9} = \frac{1}{3}$

2 **a** 7 **b** 24 **c** 32 **d** 10

e 12 **f** 5 **g** 12 **h** 9

3 $\frac{1}{3}$ of 24 → 8

$\frac{1}{4}$ of 12 → 3

$\frac{1}{10}$ of 60 → 6

$\frac{2}{5}$ of 25 → 10

$\frac{3}{4}$ of 20 → 15

4 **a** $\frac{4}{35}$ **b** $\frac{5}{12}$ **c** $\frac{3}{20}$ **d** $\frac{16}{15}$ **e** $\frac{21}{20}$

f $\frac{5}{8} \times \frac{5}{3} = \frac{25}{24}$ **g** $\frac{4}{7} \times \frac{5}{3} = \frac{20}{21}$ **h** $\frac{5}{6} \times \frac{9}{8} = \frac{45}{48}$ **i** $\frac{7}{8} \times \frac{5}{4} = \frac{35}{32}$

5 **a** $8 \times \frac{5}{3} = \frac{40}{3}$ **b** $2 \times \frac{9}{5} = \frac{18}{5}$ **c** $3 \times \frac{7}{4} = \frac{21}{4}$ **d** $6 \times \frac{8}{5} = \frac{48}{5}$

e $11 \times \frac{4}{3} = \frac{44}{3}$ **f** $5 \times \frac{7}{3} = \frac{35}{3}$ **g** $9 \times \frac{8}{7} = \frac{72}{7}$ **h** $10 \times \frac{12}{11} = \frac{120}{11}$

6 **a** $\frac{3}{10}$ **b** $\frac{4}{15}$ **c** $\frac{6}{50}$ **d** $\frac{15}{64}$

e $\frac{5}{18}$ **f** $\frac{6}{20}$ **g** $\frac{14}{24}$ **h** $\frac{27}{50}$

7 $\frac{2}{3}$ of $\frac{5}{8} = \frac{10}{24}$, which is bigger than $\frac{2}{5}$ of $\frac{3}{8}$ ($\frac{6}{40}$).

8 **a** $\frac{3}{20}$ **b** 840

9 $120 \times \frac{3}{4} = 90$ m, so she has enough wire.

10 David is correct as $\frac{1}{4} \times \frac{1}{3} = \frac{1}{3} \times \frac{1}{4} = \frac{1}{12}$.

Answers to Problem solving – The 2016 Olympic Games in Rio

1 17 days

2 **a** nearly 5250 **b** almost 7875 **c** about 2625

3 $\frac{1}{4}$

4 $\frac{9}{34}$

5 **a** **i** 18 **ii** 12 **iii** 3

b $\frac{1}{12}$

6 $216 million

7 $\frac{1}{3}$

8 3 300 000

8 Algebra

Learning objectives

- More about expanding brackets and factorising algebraic expressions
- How to simplify more complicated expressions

Prior knowledge

- How to collect like terms in an expression
- How to multiply out a simple bracket
- How to simplify simple expressions

Context

- This chapter recalls previous work on algebra and revisits expansion of brackets and collecting like terms. Pupils are also shown how to expand a bracket, and then factorise a bracket that involves powers. Finally, pupils learn how to expand expressions with two brackets.

Discussion points

- Explain how you would expand: $x(x + 8)$.
- What about $-x(x + 8)$?
- What does 'factorise as much as possible' mean?
- Give me an example where an expression has *not* been factorised as much as possible.
- Explain your method for expanding two brackets.

Associated Collins ICT resources

- Chapter 8 interactive activities on Collins Connect online platform
- *Random digits* Wonder of Maths on Collins Connect online platform

Curriculum references

Algebra

- Simplify and manipulate algebraic expressions to maintain equivalence by:
  - collecting like terms
  - multiplying a single term over a bracket
  - taking out common factors
  - expanding products of two or more binomials

Fast-track for classes following a 2-year scheme of work

- You could fast-track pupils who grasp the material in this chapter quickly to the more challenging questions at the end of each exercise in the Pupil Book.

Lesson 8.1 Expanding brackets

Learning objective

- To multiply out brackets with a variable outside them

Resources and homework

- Pupil Book 3.1, pages 129–131
- Homework Book 3, section 8.1
- Online homework 8.1, questions 1–10

Links to other subjects

- **Science** – to manipulate scientific expressions

Key words

- No new key words for this topic

Problem solving and reasoning help

- **MR** question 10 in Exercise 8A of the Pupil Book requires pupils to explain the correct process of collecting like terms. Model some examples containing numbers, if pupils fail to grasp the concept.

Common misconceptions and remediation

- Pupils occasionally fail to multiply the second term inside the bracket by the term outside the bracket. For example, pupils may get the answer: $4(x + 2) = 4x + 2$ instead of: $4x + 8$.

Probing questions

- Pupils will benefit from identifying and correcting errors when expanding expressions in the following, for example: $3(x +2) = 3x + 2$ $\quad 4(y – 6) = 4y – 10$ $\quad x(5 – 2y) = 5x – 2y$

Part 1

- Display a target board, as shown, or of your own design. Using mini whiteboards, ask pupils to answer questions such as:
  - Which numbers are square numbers?
  - Which are prime numbers?
  - Add the four corner numbers together.
  - What is the sum of the first column minus the sum of the second column?

8	5	15	4	7
3	12	9	11	20
10	13	100	2	3

Part 2

- On the board, write 3(2 + 5). Ask the class for the value. After a short discussion about pupils' answers, draw the rectangle as shown.

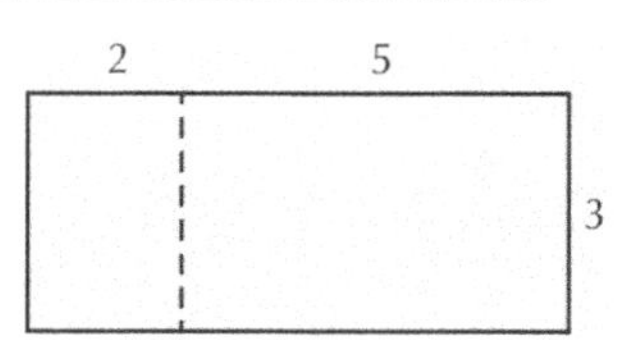

- Show how the large 7 × 3 rectangle has been split into two smaller rectangles, 2 × 3 and 5 × 3. Use this illustration to show that:

  $3(2 + 5) = 3 \times 2$ and 3×5

  $= 6 + 15 = 21$.

- Draw the rectangle as shown on the right. Show the area as $2(3 + x)$, which is made up of $2 \times 3 + 2 \times x$. This is $6 + 2x$, so $2(3 + x) = 6 + 2x$. Discuss expanding this bracket.

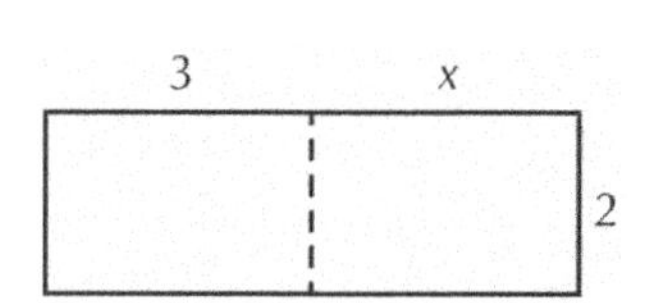

- Next, write the expression $4(2x + 3)$. Ask pupils what this represents. You want the answer to be a rectangle of sides 4 and $2x + 3$.
- Then ask pupils what the expansion of $3(5 + 2x)$ gives. $(15 + 6x)$
- On the board, draw the rectangle as shown on the right. Ask pupils to give you the area of the shape. The area is given by $m(4p + 2)$ or $4mp + 2m$. Emphasise that this illustrates that $m(4p + 2) = 4mp + 2m$.

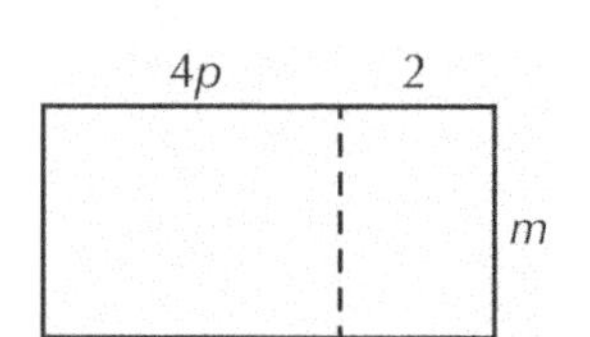

- Tell the class that they have been looking at rectangles to help them understand how the expansion of brackets works. However, pupils do need to see brackets with negative multipliers before they try to expand some themselves.
- On the board, write $5(4k - 3)$. Ask for suggestions. Discussion may lead to $20k - 15$. Some pupils may point to a rectangle with an end cut off. Discuss this if it is suggested.
- Next, give the class $t(2t - 4)$ and proceed as before, ending with $2t^2 - 4t$.
- **Pupils can now do Exercise 8A from Pupil Book 3.1.**

Part 3

- On the board, write: 2( ) = $6 + 8m$. Ask the class what might be inside the bracket.
- This will lead to $3 + 4m$, but discuss with pupils how they knew.
- Write another expression to be solved: m( ) = $3m + mt$. Discuss what might be inside the bracket. $(3 + t)$

Answers

Exercise 8A

1	**a** $7x$	**b** $10a$	**c** $9t$	**d** $9y$	
	e $6m$	**f** $4k$	**g** $6n$	**h** $-4p$	
2	**a** $10m$	**b** $9y$	**c** $12t$	**d** $14p$	
	e $14n$	**f** $9p$	**g** $11t$	**h** $6e$	
	i $6k$	**j** $3h$	**k** $2m$	**l** $6t$	
3	**a** $9x$	b $12a$	**c** $20t$	**d** $12y$	**e** $15k$
	f $12t$	**g** $21x$	**h** $10m$	**i** $24t$	**j** $24y$
4	**a** $4t + 12$	**b** $2x + 12$	**c** $16m - 8$	**d** $10k - 15$	
	e $9 + 6x$	**f** $20 - 12k$	**g** $14 - 6y$	**h** $15 - 3x$	
5	**a** $xy + 2x$	**b** $3am + 2m$	**c** $2kp + 4k$	**d** $6mn + 3n$	
	e $5t + 4qt$	**f** $3g + 4gh$	**g** $7h + 5gh$	**h** $3k + 2dk$	
	i $4ab - 3a$	**j** $5c - 4cd$	**k** $2f - 3fm$	**l** $5b - 4ab$	
	m $5ad + 3d$	**n** $7ef + 3e$	**o** $3xy + 2y$	**p** $2pq + 5p$	
	q $3q - 4pq$	**r** $6t - 3st$	**s** $8w - 5kw$	**t** $3n - 2mn$	
6	**a** $A = xy + 5y$	**b** $A = 2mx + 3m$	**c** $A = 6d + 3ad$		
	d $A = 2ak + 3k$	**e** $A = 3n + 5ny$	**f** $A = 5pq + 6q$		
7	**a** $4x^2$	**b** $5a^2$	**c** $6t^2$	**d** $4y^2$	**e** $2k^2$
	f $5t^2$	**g** $8x^2$	**h** $3m^2$	**i** $4t^2$	**j** $5y^2$
8	**a** $x^2 + 2x$	**b** $3m^2 + 2m$	**c** $4k^2 + k$	**d** $4n^2 + 3n$	
	e $6t + 2t^2$	**f** $g + 4g^2$	**g** $3h + 5h^2$	**h** $2d + 3d^2$	
	i $5a^2 - 2a$	**j** $3c - 4c^2$	**k** $5t - 3t^2$	**l** $7b - 4b^2$	
	m $8d^2 + 7ad$	**n** $5e^2 + 3e$	**o** $2xy + 3y^2$	**p** $5p + 4p^2$	
	q $7q^2 - 5q$	**r** $2t^2 - 5t$	**s** $3w^2 - 4w$	**t** $8n^2 - 5n$	
9	**a** $A = 4m^2 + 3m$	**b** $A = 6t + 3t^2$	**c** $A = 3k^2 + k$		
	d $A = 4x + 3x^2$	**e** $A = 2g^2 + 7g$	**f** $A = 3n + 2n^2$		

10 $3x^2$ and $5x$ are not like terms, so cannot be added together

Challenge: Mixed letters

A 9

B $x - 1$

Lesson 8.2 Factorising algebraic expressions

Learning objective

- To factorise expressions

Resources and homework

- Pupil Book 3.1, pages 132–133
- Homework Book 3, section 8.2
- Online homework 8.2, questions 1–10

Links to other subjects

- **Science** – to manipulate scientific expressions

Key words

- No new key words for this topic

Problem solving and reasoning help

- The **PS** and **MR** questions at the end of Exercise 8B in the Pupil Book require pupils to apply their learning from this lesson and from Lesson 8.1 of this chapter. Before pupils work on the investigation at the end of the exercise, review expanding brackets and collecting like terms.

Common misconceptions and remediation

- Pupils often fail to fully factorise. For example, $x^2 + x^3y$ pupils may be factorised incorrectly as $x(x + x^2y)$. Remind pupils that they need to take care that they factorise expressions fully.

Probing questions

- Pupils will benefit from identifying and correcting errors when factorising expressions, for example those in $12x + 6 = 3(4x +2)$ and $4y^2 - 12y^3 = 2y(y - 6y^2)$.

Part 1

- Show pupils the following bracket expansions and ask them to spot any errors
  - $2(x + 4) = 2x + 4$
  - $x(x + y) = x + xy$
  - $y(3 + w) = y^3 + wy$
  - $x(4x - x) = 4x^2 + x^2$
  - $2(x + 4) = 2x + 4$
  - $x(3 - x^2) = 3x - x^2$

Part 2

- Remind pupils that factorising is the opposite of multiplying out (or expanding) brackets. Explain that when multiplying out brackets, the expression $4(x + y)$ can be expanded to $4x + 4y$. Factorising $4x + 4y$ produces $4(x + y)$, which is the original expression.
- Go through the following examples with pupils:
  - Factorise $3a + 12$:
    3 is a factor of both $3a$ and 12, so place this outside a pair of empty brackets 3(   )
  - Decide what you have to multiply by 3 to produce the original expression
    That is, $3 \times a = 3a$ and $3 \times 4 = 12$, so: $3a + 12 = 3(a + 4)$
  - Explain how to check this answer by multiplying out $3(a + 4)$.
    This gives $3a + 12$, which is the original expression.
  - Factorise ab + 2a:
    a is a factor of both ab and $2a$, so place this outside a pair of empty brackets a(   )
  - Decide what you have to multiply by a to produce the original expression
    That is, $a \times b = ab$ and $a \times 2 = 2a$, so: $ab + 2a = a(b + 2)$
  - Explain how to check this answer by multiplying out $a(b + 2)$.
    This gives $ab + 2a$, which is the original expression.
- **Pupils can now do Exercise 8B from Pupil Book 3.1.**

Part 3

- Reinforce *factorisation* by asking pupils to exlain what it means. Responses should show an understanding of breaking down an expression into two terms that will multiply together to give the original. This explanation can be used as a literacy task.

Answers

Exercise 8B

1 **a** 1, 2, 3, 4, 6, 8, 12, 24 **b** 1, 5, 7, 35 **c** 1, 2, 4, 5, 8, 10, 20, 40
d 1, 2, 4, 7, 14, 28 **e** 1, 2, 3, 4, 6, 9, 12, 18, 36 **f** 1, 2, 3, 6, 9, 18
g 1, 2, 4, 7, 14, 28 **h** 1, 2, 5, 10, 25, 50

2 **a** 1, 2, x, $2x$ **b** 1, 3, m, $3m$ **c** 1, 2, 4, t, $2t$, $4t$ **d** 1, 5, y, $5y$
e 1, 3, x, $3x$, x^2, $3x^2$ **f** 1, 2, m, $2m$, m^2, $2m^2$ **g** 1, 5, t, $5t$, t^2, $5t^2$ **h** 1, 3, k, $3k$

3 **a** t **b** q **c** x **d** a
e t **f** q **g** x **h** a

4 **a** $m(x + 2)$ **b** $m(t + 3)$ **c** $p(n + 2)$
d $x(x - 3)$ **e** $p(p - 1)$ **f** $y(y - 2)$
g $k(4 + x)$ **h** $k(3 + k)$ **i** $x(2 - t)$

5 **a** $t(3 + m)$ **b** $x(2 + y)$ **c** $p(5 + q)$
d $k(6 - k)$ **e** $n(n - 5)$ **f** $x(x - 8)$
g $x(5 + x)$ **h** $h(1 + h)$ **i** $t(2 - 3t)$

6 **a** x **b** p **c** $t + 5$ **d** $4 - y$

7 **a** $x^2 + 5x = x(x + 5)$ **b** $6m + mt = m(6 + t)$ **c** $3y^2 + 2y = y(3y + 2)$

8 **a** $t(4t + 5)$ **b** $x(6x + 1)$ **c** $2(3t + 2)$
d $x(3x - m)$ **e** $t(5t + k)$ **f** $3(3x + 2)$
g $t(5 + 8t)$ **h** $x(3x - 2)$ **i** $5(2t + 3)$

9 For example, find the highest common factor, which is p, and write this outside the brackets. Then divide $12p^2$ and $5p$ by p to get the terms in the brackets.
So $12p^2 + 5p = p(12p + 5)$.

Investigation: An age-old problem

Andrew's grandpa was 27 when Andrew's dad was born.

Lesson 8.3 Expand and simplify

Learning objectives

- To expand expressions with two brackets and simplify them

Resources

- Pupil Book 3.1, pages 133–135
- Homework Book 3, section 8.3
- Online homework 8.3, questions 1–10

Links to other subjects

- **Science** – to manipulate scientific expressions

Key words

- No new key words for this topic

Problem solving and reasoning help

- The questions in Exercise 8C of the Pupil Book relate the expansion of brackets back to models involving the area of a rectangle. These were introduced in Lesson 8.1, but it may be beneficial to go over them with pupils in the context of multiplying out two brackets before they tackle the questions in the exercise.

Common misconceptions and remediation

- Pupils will sometimes make mistakes when there is a negative sign outside the bracket. Model several examples, making sure pupils remember the correct procedures.

Probing questions

- Explain how you would multiply out an expression containing two brackets.
- Pupils will benefit from identifying and correcting errors when expanding and simplifying expressions, for example those in:

  $5(2x + 4) -2 (3x - 4) = 10x + 20 - 6x - 8 = 16x + 12$ (spot the errors)

Part 1

- Set percentages in context by talking about getting a 20% reduction on the price of a certain item you purchased recently.
- Ask pupils how they would calculate 20% of a value. Pupils might use the strategy of finding the value of 10%, then doubling it, or they might divide the original value by 5.
- Give an example such as: 'What is 20% of £34?' Work through the example using the first method, which most pupils will find is more straightforward. That is, 10% of £34 is £3.40; doubling it gives 20%, as £6.80.
- Using a target board such as this one, work your way around the class, asking individual pupils to work out 20% of each given value.

£45	35 kg	50 minutes	£82	29 kg
65 minutes	£67	18 metres	2 hours	£39
130 kg	180 minutes	£89	75 kg	49 metres
£234	83 metres	24 hours	£26	130 kg

Part 2

- This lesson is concerned mainly with making algebraic expressions as simple as possible.
- Write on the board $4(5 + 2y) + 2(5y - 6)$. Ask for someone to expand the first bracket. Then ask for someone to expand the second bracket.
- You will end up with $20 + 8y$ and $10y - 12$ on the board. Referring to the original expression, ask the class: 'Will adding these together make the result simpler?'
- The answer is, 'Yes', but first the like terms, $8y$ and $10y$, 20 and –12, must be gathered together. This gives $8y + 10y = 18y$ and $20 - 12 = 8$. That is, $18y + 8$ is the simplified form.
- Now work through another example with the class: $x(3x + 4) - x(x - 5)$.

Expand this to $3x^2 + 4x - x^2 + 5x$.

- Remind pupils that two negative values multiplied together give a positive result. Hence the last term is $-x \times -5 = +5x$.
- Now bring the like terms together: $3x^2 - x^2 + 4x + 5x$, giving $2x^2 + 9x$.
- **Pupils can now do Exercise 8C from Pupil Book 3.1.**

Part 3

- Working in pairs, challenge pupils to expand and simplify: $3(4x + 6) - 2(10 + 5x) + 2(1 - x)$.
- The answer is 0, as the *x*s and the numbers cancel out each other. See who will be first to get the correct answer. Ask pupils not to shout out the answer but to tell you quietly.

Answers

Exercise 8C

1	**a** $10m$	**b** $12y$	**c** $8t$	**d** $16p$
	e $12n$	**f** $13p$	**g** $8t$	**h** $4e$
	i 0	**j** $-h$	**k** $-2m$	**l** $6t$
2	**a** $10t + 8g$	**b** $9x + 5y$	**c** $6m + 4k$	**d** $6x + 8y$
	e $3m + 3p$	**f** $4n + 9t$	**g** $9k + 2g$	**h** $3d + 6b$
	i $3q + 2p$	**j** $8g + 2k$	**k** $8x - 6y$	**l** $2e - 4d$
3	**a** $4y + 12$	**b** $6a + 8$	**c** $10p + 15$	**d** $6m + 9$
	e $4t + 3qt$	**f** $2g + 5gh$	**g** $3h + 7gh$	**h** $4k + 3dk$
	i $3a^2 - 3a$	**j** $4c - c^2$	**k** $2f - 3f^2$	**l** $5b - 4b^2$
	m $15a + 10$	**n** $5ef + e$	**o** $y^2 + 4y$	**p** $2p^2 + 3p$
4	**a** $9x + 14$	**b** $20k + 33$	**c** $16t + 23$	**d** $18q + 11$
	e $26h + 8$	**f** $34 + 9f$	**g** $18 + y$	**h** $27t - 36$
5	**a** $4x + 9$	**b** $2k + 18$	**c** $12t + 9$	**d** $2q + 5$
	e $4h + 46$	**f** $w + 37$	**g** $11x - 9$	**h** $6t - 21$
6	**a** $6x^2 + 8x$	**b** $5p^2 + 5p$	**c** $7k^2 + 7k$	**d** $5d^2 + 8d$
	e $8n^2 + n$	**f** $8f^2 + f$	**g** $3p^2 - 9p$	**h** $9y^2 - 5y$
7	**a** $4x^2 + 4x$	**b** $3p^2 + 3p$	**c** $2k^2 + k$	**d** $3f^2 + 9f$
	e $4n^2 + 7n$	**f** $3f^2 + 9f$	**g** $3p^2 + 4p$	**h** $2y^2 + 4y$

8 yes, $3(x + 5) + x(x + 5) = 3x + 15 + x^2 + 5x = x^2 + 8x + 15$

Challenge: All legs and heads

There are 22 more cows (36) than chickens (14).

Review questions (Pupil Book pages 136–137)

- The review questions will help to determine pupils' abilities with regard to the material within Chapter 8.
- These questions also draw on the mathematics covered in earlier chapters of the book to encourage pupils to make links between different topics.
- The answers are on the next page of this Teacher Pack.

Challenge – California Gold (Pupil Book pages 138–139)

- Ask pupils if they know what treasure trove means.
- Explain that it is an amount of money or coin, gold or silver, hidden underground or in places such as cellars or attics, where the treasure seems old enough for people to assume that the true owner is dead and the heir or heirs are untraceable.
- Pupils could research 'largest treasure trove finds' and 'most valuable coins' on the internet. Pupils could also research the unit of measurement for gold (1 troy ounce = 31 grams).
- Now go through the information relating to the coins found in the Pupil Book. Ask pupils questions such as:
    - How old are the coins?
    - How much would three coins be worth?
    - How much would x coins be worth?
    - How many zeros does the number billion have in it?
    - Gold in 2014 is worth £773 per troy ounce. How much is 1 gram worth?
- Pupils can now answer the questions in the Pupil Book.

Answers to Review questions

1	**a** $8p$	**b** $13x$	**c** $12q$	**d** $9t$
	e $10n$	**f** $10p$	**g** $6m$	**h** $4a$
	i h	**j** $-2g$	**k** n	**l** $-3t$
2	**a** $8x + 3$	**b** $12t - 1$	**c** $10t + 13$	
3	**a** $2m + 8$	**b** $4t + 20$	**c** $3x - 21$	**d** $5t - 10$
	e $3m + my$	**f** $3t - th$	**g** $5x - 2tx$	**h** $2k - 3kt$
4	**a** $A = 3t + 12$	**b** $A = tx - 3t$	**c** $A = m^2 + 4m$	**d** $A = t^2 + 3t$
5	**a** $2(3t + 5)$	**b** $3(x + 2)$	**c** $t(5 - 2m)$	
	d $x(6 - y)$	**e** $t(4 - 5p)$	**f** $x(3 + m)$	
	g $t(5 + 3)$	**h** $x(2x + 7)$	**i** $t(4t - \mathrm{m})$	
6	**a** t	**b** $x + 4$	**c** $m - p$	

7 **a** $t(t - 8) = t^2 - 8t$
The correct factorisation is $t(8 - t)$.
b The common factor is p, not m.
The correct factorisation is $p(3m + 2p)$.
c This one is correct.

8	**a** $11x + 23$	**b** $14k + 21$	**c** $18t + 16$	**d** $13q + 21$
	e $2h + 13$	**f** $-w + 18$	**g** $6x - 2$	**h** $2t + 8$
9	**a** $8x^2 + 8x$	**b** $9p^2 + 5p$	**c** $7k^2 + 7k$	**d** $5d^2 + 8d$
	e $3n^2 + 2n$	**f** $3f^2 + 7f$	**g** $p^2 + p$	**h** $2y^2 + 3y$

10 Find the common factor, which is t, and write this outside the brackets. Then work out what you need to multiply this by to get $5t^2 - 2t$. So $t(5t - 2)$.

Answers to Challenge – California Gold

1	**a** 128 years	**b** 167 years	
2	**a** 30 grams	**b** $15x$ grams	**c** 21 405 grams (21.4 kg)
3	**a** £56	**b** £28y	**c** £599 340
4	**a** £1 200 000	**b** £600 000w	**c** £856 200 000

5 Mrs T is correct, as the total value could be £0.856 billion.

6 The coins are worth more as they are. Each coin is worth more than the total value as gold.

9 Decimal numbers

Learning objectives

- How to extend your ability to work with powers of 10
- How to know when to make suitable rounding and to use rounded numbers to estimate the results of calculations

Prior knowledge

- How to multiply and divide by 10 and 100

Context

- The ability to understand place value is the key to being able to use numbers effectively when doing calculations in real life. Share with pupils some of the different aids to calculation that have been developed since we first used numbers to count. For example, skilled abacus users can calculate very quickly and accurately. You could show a video such as the one at the following link (about Japanese children learning super-fast mathematics with the abacus) to demonstrate different approaches to calculations across the world:
  http://www.youtube.com/watch?v=6m6s-uIE6LY
- Nowadays, computers and calculators can help you to work out the more complicated calculations. Modern calculators can do so much more than simple arithmetic, but pupils need to know how to integrate this with a good basic understanding of place value and mental arithmetic. In this chapter, pupils will learn more about the decimal system of counting and they will practise skills in using a calculator.

Discussion points

- My calculator display shows 0.001. Tell me what will happen when I multiply by 100. What will the display show?
- I divide a number by 10, and then again by 10. The answer is 0.3. What number did I start with? How do you know?
- How would you explain how to multiply a decimal by 10, and how to divide a decimal by 100?
- What is the same and what is different about these numbers: 84.544 and 84.547?

Associated Collins ICT resources

- Chapter 9 interactive activities on Collins Connect online platform
- *Sporting decimals* Wonder of Maths on Collins Connect online platform

Curriculum references

Develop fluency

- Consolidate their numerical and mathematical capability from Key Stage 2 and extend their understanding of the number system and place value
- Select and use appropriate calculation strategies to solve increasingly complex problems

Number

- Understand and use place value for decimals, measures and integers of any size
- Round numbers and measures to an appropriate degree of accuracy

Fast-track for classes following a 2-year scheme of work

- Pupils should have met most of the material in this chapter before. However, this material may challenge some pupils. It is important to remember that lack of confidence and fluency with basic number skills can be a significant barrier to further learning for pupils working at this level. However, if you feel your pupils are able to move on faster, you could combine Lesson 9.1 and Lesson 9.4, and then Lesson 9.2 and Lesson 9.3, by choosing key questions in each pair of lessons. Then move on to Lesson 9.5.

Lesson 9.1 Multiplication of decimals

Learning objective

- To practise multiplying decimal numbers

Resources

- Pupil Book 3.1, pages 141–143
- Intervention Workbook 2, pages 20–21
- Intervention Workbook 3, pages 6–8
- Online homework 9.1, questions 1–10

Links to other subjects

- **Science** – to check whether or not calculations as part of experimental results make sense
- **Design and technology** – to estimate the amount of raw material required for a product

Key words

- No new key words for this topic

Problem solving and reasoning help

- **PS** questions 8, 9 and 10 of Exercise 9A in the Pupil Book apply pupils' learning to real-life contexts involving area and money.

Common misconceptions and remediation

- Some pupils treat the numbers as whole numbers, (ignoring the decimal point). These pupils then simply line up the decimal point back in a similar way as they would for addition. Emphasise an understanding of place value and estimating answers before trying to do the calculation, which may help pupils to see for themselves the error of doing it this way. Use examples such as 0.5 × 0.5 and use language such as: What is half of a half?

Probing questions

- Explain how you would do this multiplication by using factors such as: 6.4 × 50. What clues do you look for when deciding if you can do a multiplication involving decimals mentally?

Part 1

- Prepare some cards with large whole numbers on them, for example: 200, 30, 4000.
- Produce two cards and ask pupils to multiply the numbers together.
- Check pupils' answers, then do another example (or more examples).

Part 2

- This is a revision lesson on multiplying decimals.
- Recall the rules for making calculations such as 0.3 × 0.05. Start with 3 × 5 = 15 and note that there are 0._ × 0._ _ = 0._ _ _ three decimal places, so the answer is 0.015.
- Give some examples, such as: 0.004 × 0.03 (= 0.000 12) and 0.5 × 0.007 (= 0.0035)
- Recall the rules for making calculations such as 200 × 0.007. Rewrite as equivalent products until suitably simplified. For example: 200 × 0.007 = 20 × 0.07 = 2 × 0.7 = 1.4
- Give examples, such as:
  300 × 0.07 = 30 × 0.7 = 3 × 7 = 21; 40 × 0.0008 = 4 × 0.008 = 0.032
- Confirm the principle that to maintain equality when multiplying two numbers, when one number is multiplied by 10, the other number must be divided by 10.
- Work through Example 3 on page 142 of the Pupil Book to demonstrate long multiplication.
- **Pupils can now do Exercise 9A from Pupil Book 3.1.**

Part 3

- Discuss the reason we count in a base of 10 – probably because we have 10 fingers.
- Ask pupils to name as many words beginning with 'dec-' that relate to 10. For example: decagon, decade, decimate, December, decahedron, decalitre, decalogue, decametre, decapod, decastyle, decasyllabic, decathlon, decennial, decibel, decilitre, decimetre.
- You could ask pupils to find as many as possible before the next lesson.

Answers

Exercise 9A

1	**a** 199.2	**b** 199.2	**c** 19.92	**d** 0.1992	
2	**a** 80	**b** 80	**c** 8	**d** 0.8	
3	**a** 13	**b** 20.4	**c** 19.64	**d** 30.6	
	e 220.5	**f** 179.2	**g** 87.6	**h** 76.4	
4	**a** 5	**b** 7	**c** 30	**d** 6	
	e 7	**f** 80	**g** 10	**h** 40	
5	**a** 20	**b** 14	**c** 60	**d** 30	
	e 28	**f** 240	**g** 40	**h** 200	
6	**a** 0.18	**b** 0.25	**c** 0.63	**d** 0.36	
	e 0.72	**f** 0.42	**g** 0.4	**h** 0.16	
	i 0.49	**j** 0.27	**k** 0.32	**l** 0.06	
7	**a** 492	**b i** 0.492	**ii** 0.492		
8	**a** 1.26 m²	**b** 2.53 m²			
9	£20.60				
10	£187.50				

Investigation: Mystical multiplication

A	**a** 1	**b** 0.49	**c** 0.09	**d** 0.4	**e** 0.4

B They are the same.

C pupils' own work

D In each case **d** and **e** give the same answer. When $x + y = 1$, then $x^2 - y^2 = x - y$.

Lesson 9.2 Powers of 10

Learning objective

- To understand and work with both positive and negative powers of ten

Resources

- Pupil Book 3.1, pages 143–147
- Homework Book 3, section 9.1
- Online homework 9.2, questions 1–10

Links to other subjects

- **Science** – to understand the physical and biological world using very large and very small numbers
- **History** – to use an extended historical timescale

Key words

- negative power

Problem solving and reasoning help

- **PS** question 12 in Exercise 9B of the Pupil Book uses population figures, which are a natural example of how large numbers can get in reality. You could introduce the activities with a familiar example such as the population of a local town. The activity at the end of Exercise 9B makes links to standard units of measure.

Common misconceptions and remediation

- Place value is a key concept in mathematics. The work in this chapter builds on pupils' existing knowledge. If necessary, check earlier objectives involving an understanding of place value to consolidate this understanding and prepare pupils for this lesson. Most pupils will understand that each column to the left of another is 10 times greater. Build on this so that pupils are aware that each column to the right is 10 times smaller.

Probing questions

- How would you explain that 0.35 is greater than 0.035?
- Why do these give the same answer: 25 ÷ 10 and 250 ÷ 100?
- My calculator display shows 0.001. What will the display show when I multiply by 100?
- I divide a number by 10, and then again by 10. The answer is 0.3. What number did I start with? How do you know?

Part 1

- Have prepared cards available of the two calculator displays, or write them on the board. Ask pupils to write the numbers in full on mini whiteboards or in their exercise books.

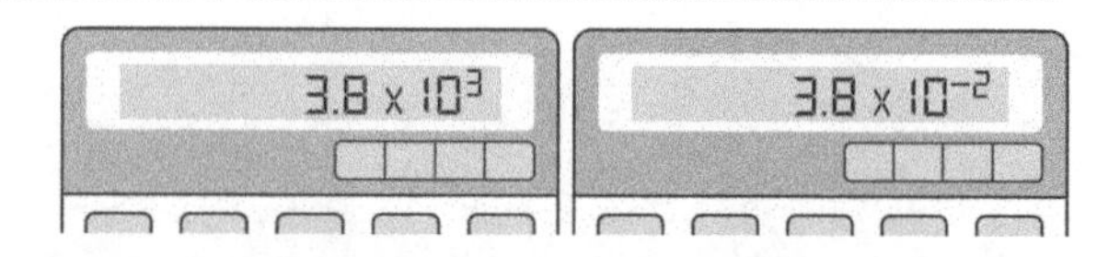

- You may need to explain that the power gives the number of places to move the digits. Negative powers mean move to the right, positive powers mean move to the left. Pupils may prefer to 'see' the decimal point shifting. Say that on a calculator, a negative power moves the point to the left and a positive power moves the point to the right.
- Check pupils' answers and then call out (or display) the answers.
- If pupils are writing in exercise books, give 10 or more examples before checking answers.

Part 2

- This lesson recalls working with powers of 10 and follows on from Part 1. Write a problem on the board, for example, 6.3 × 100. Ask pupils for the answer (630), and to explain how they worked it out. (They moved the decimal point.) In the context of this lesson, this may be the best way for pupils to see what is happening, but remind them that the digits will move.
- Write another problem on the board such as 7.8×10^2. Ask for the answer (780), and what connection there is with the previous example. (The power and the number of zeros give the number of places that the digits move.) Repeat with 0.32 ÷ 1000 and $0.32 \div 10^3$. (0.000 32)
- Write a third problem on the board, for example, 67.2 × 0.01. Ask pupils for the answer (0.672), and to explain how they worked it out. Discuss any similarities with previous examples. Pupils may realise that multiplying by 0.01 is the same as dividing by 100 or 10^2.
- **Pupils can now do Exercise 9B from Pupil Book 3.1.**

Part 3

- Reverse Part 1: Write, for example, 640 on the board. Ask pupils to write it as a calculator display: 6.4×10^2
- Repeat with other numbers such as: 0.0067, 8900, 0.53, 510 000.

Answers

Exercise 9B

1	**a** 570	**b** 6900	**c** 78 000		
	d 714 000	**e** 80 200	**f** 3150		
2	**a** 74.8	**b** 3.29	**c** 0.473		
	d 0.058	**e** 0.85	**f** 1.7		
3	**a** 143	**b** 3620	**c** 57 300		
	d 32 140	**e** 1285	**f** 391.7		
4	**a** 0.634	**b** 0.473	**c** 0.0663		
	d 0.0027	**e** 0.0376	**f** 7.193		
5	**a** 115	**b** 0.637	**c** 42 300		
	d 0.003 65	**e** 107	**f** 0.914		
	g 41	**h** 0.038	**i** 7400		
6	**a** 0.1	**b** 0.01	**c** 0.001		
	d 0.000 01	**e** 0.000 001			
7	**a** 10^{-2}	**b** 10^{-3}	**c** 10^{-1}	**d** 10^{-4}	**e** 10^{-7}
8	**a** 960	**b** 18	**c** 20 460	**d** 1297	
9	**a** 8160	**b** 710	**c** 824 600	**d** 29 660	
10	**a** 0.77	**b** 0.063	**c** 51.46	**d** 23.58	
11	**a** 0.186	**b** 0.0184	**c** 2.185	**d** 0.346	

12

	Population	
Singapore	Five million	$14.2 \times 10^{-5} = 0.000\ 142$
Hong Kong	Seven million	$15.7 \times 10^{-5} = 0.000\ 157$
Belgium	Eleven million	$27.3 \times 10^{-4} = 0.00\ 273$
Japan	127 million	$29.9 \times 10^{-4} = 0.00\ 299$

Activity: Prefixes

A Use suitable prefixes to write each quantity in a simpler form.

a 3 megawatts **b** 5 kilometres **c** 3 gigabytes

d 7 centigrams **e** 4 millimetres **f** 5.5 microlitres

B approximately 30 centimetres

Lesson 9.3 Rounding suitably

Learning objective

- To round numbers, where necessary, to a suitable degree of accuracy

Resources

- Pupil Book 3.1, pages 147–149
- Intervention Workbook 3, pages 9–10
- Homework Book 3, section 9.3
- Online homework 9.3, questions 1–10

Links to other subjects

- **Physical education** – to give sporting results to an appropriate degree of accuracy
- **Science** – to give experimental results to an appropriate degree of accuracy
- **Design technology** – to give measurements to an appropriate degree of accuracy

Key words

- suitable degree of accuracy

Problem solving and reasoning help

- The rounding activity at the end of Exercise 9C in the Pupil Book uses a matching exercise to connect rounding with estimation. Ask pupils for examples in real life. You could also ask pupils to talk to family and friends about when they use rounding and estimation.

Common misconceptions and remediation

- Pupils often find decimal fractions difficult to understand, yet they may have used them when working with money or solving money problems. Working with decimals is an extension of pupils' understanding of place value. Most pupils understand that each column to the left is 10 times greater; they also need to grasp that each column to the right is 10 times smaller.

Probing questions

- Are the following true or false? Explain your answers.
    - 4.299 rounds to 4.210 to two decimal places.
    - 3.5 is closer to 4 than it is to 3.
    - 8.4999 rounds to 8.5 to one decimal place.
- How do you go about rounding a number to one decimal place?
- Explain whether the following is true or false: 10 is greater than 9, so 0.10 is greater than 0.9.

Part 1

- Using a target board as shown, ask pupils to round the numbers so that only one non-zero digit plus some zeros, if necessary, remain. You may need to demonstrate this to pupils by giving them a few examples.

4.652	0.0752	1.071	3.222	0.578
0.0825	1.629	11.635	3.999	4.814
3.421	8.525	3.688	9.002	1.035
6.455	1.459	1.291	5.927	2.716

  For example, the answers for the top row are: 5, 0.08, 1, 3, 0.6.
- Then ask groups or individuals to round the remaining numbers to one significant figure (1 sf). Use this terminology if it is appropriate.
- Next, repeat with rounding so that two non-zero digits (and zeros) remain. Again, you may need to provide examples. The answers for the top row are: 4.7, 0.075, 1.1, 3.2, 0.58.

Part 2

- This lesson builds on rounding, and makes links to some of the other ideas in this chapter.
- Remind pupils of the two main reasons for rounding: one is to answer a question to an appropriate degree of accuracy; the other is to make an estimate of the answer to a problem.
- Start by rounding some numbers to one and two decimal places. For example (give more if necessary): 2.367 = 2.4 (1 dp) = 2.37 (2 dp); 0.825 = 0.8 (1 dp) = 0.83 (2 dp).
- Now ask pupils to round 56 to the nearest number with only one non-zero digit and one or more zeros, as in Part 1 (60). Repeat with 489 (500); 0.31 (0.3); 0.0065 (0.007); 0.99 (1).
- Introduce, if not already mentioned in Part 1, the terminology *one significant figure* (1 sf).
- Ask for an estimate of 32% of £529. Pupils may guess, but ask for a method. For example, round to 1 sf to give: 32% of £529 ≈ 30% of £500 = 3 × 10% of £500 = 3 × £50 = £150.
- Repeat with $2.3^2 \div 0.398 \approx 2^2 \div 0.4 = 4 \div 0.4 = 40 \div 4 = 10$. Do more examples if necessary.
- **Pupils can now do Exercise 9C from Pupil Book 3.1.**

Part 3

- Ask pupils if they know the term *ballpark figure*. Obtain from pupils (or tell them) that it means an approximate value. The term comes from the USA, where the attendance at a baseball game was given approximately before accurate counting became possible.
- Ask pupils: 'If the attendance at a football match were given as 42 000 to the nearest 1000, what could the limits of the crowd be?' 41 500 and 42 499 (accept 42 500).
- Repeat with: the limits of 500 kg to the nearest 100 kg (450–550); 16 cm to the nearest cm (15.5–16.5); 1.7 to one decimal place (1.65–1.75).

Answers

Exercise 9C

1 **a** 1.3 **b** 24.2 **c** 9.0 **d** 31.1
e 1.9 **f** 5.1 **g** 2.0 **h** 4.3

2 **a** 3.26 **b** 35.19 **c** 7.04 **d** 42.14
e 2.89 **f** 6.15 **g** 3.97 **h** 7.26

3 **a i** 1.3 **ii** 1.28 **b i** 46.2 **ii** 46.17
c i 6.8 **ii** 6.84 **d i** 17.1 **ii** 17.14
e i 4.0 **ii** 4.00 **f i** 7.1 **ii** 7.07
g i 2.8 **ii** 2.75 **h i** 2.2 **ii** 2.15

4 **a** 4 + 8 = 12 **b** 8 + 2 = 10 **c** 7 – 2 = 5 **d** 7 – 3 = 4
e 6 × 8 = 48 **f** 9 × 9 = 81 **g** 8 ÷ 2 = 4 **h** 9 ÷ 3 = 3

5 **a** 30 + 70 = 100 **b** 70 + 30 = 100 **c** 80 – 30 = 50 **d** 60 – 20 = 40
e 70 × 70 = 4900 **f** 80 × 80 = 6400 **g** 60 ÷ 30 = 2 **h** 80 ÷ 20 = 4

6 **a** 300 + 100 = 400 **b** 800 + 40 = 840 **c** 800 – 100 = 700
d 500 – 80 = 420 **e** 200 × 100 = 20 000 **f** 300 × 40 = 12 000
g 200 ÷ 40 = 5 **h** 400 ÷ 20 = 5

7 Note that each answer must be accompanied by a sensible reason.
a ii, 60 mph **b** ii, 23° **c** i, 50 kg **d** ii, 4 minutes
e i, 12.8 seconds **f** ii, 2 km

Activity: Rounding

8.3 × 3.9 = 32.37, as 8 × 4 = 32
11.4 ÷ 1.5 = 7.6, as 12 ÷ 2 = 6
9.3 × 6.1 = 56.73, as 9 × 6 = 54
84 ÷ 3.2 = 26.25, as 90 ÷ 3 = 30

Lesson 9.4 Dividing decimals

Learning objective

- To confirm ability to divide with decimals

Resources

- Pupil Book 3.1, pages 150–151
- Intervention Workbook 3, pages 6–8
- Online homework 9.4, questions 1–10

Links to other subjects

- **Science** – to check whether or not calculations as part of experimental results make sense
- **Design and technology** – to estimate the amount of raw material required for a product

Key words

- No new key words for this topic

Problem solving and reasoning help

- **PS** questions 6 to 10 in Exercise 9D of the Pupil Book allow pupils to apply their learning to real-life contexts. The investigation at the end of the exercise makes links, as does Lesson 9.2. Use discussion and probing questions to make sure that pupils have made the link.

Common misconceptions and remediation

- Emphasising an understanding of place value and estimating answers before trying to do the calculation may help pupils. It may also help to use visual images and link to fractions.

Probing questions

- What clues do you look for when deciding if you can do a division involving decimals mentally?
- 26 × 57 = 1482. Explain how you can use this fact to devise calculations with answers of: 14.82 1.482 and 0.1482.

Part 1

- Write the following on the board: (14.7 × 3.9) ÷ (0.96 + 0.59).
- Ask pupils to round to 1 sf and approximate the answer: (10 × 4) ÷ (1 + 0.6) = 40 ÷ 1.6. Now the problem is too difficult to do mentally.
- Ask if it is possible to approximate differently.
- For example: (15 × 4) ÷ (1 + 0.5) = 60 ÷ 1.5 = 40. Say that approximations do not have to be made to 1 sf when a more sensible approximation is possible. Try this with 23.6 × 7.8 ÷ 49.2. Rounding to 1 sf gives 20 × 8 ÷ 50, but rounding to 25 × 8 ÷ 50 = 4 is easier to do mentally.

Part 2

- Suggest to pupils that they will often have to do calculations mentally. Give pupils a few minutes to think of some examples. Take feedback.
- Say that there is usually more than one way of doing a calculation, but that sometimes one way is more efficient than another and/or less likely to lead to mistakes.
  Work through Example 11 on page 150 of the Pupil Book.
- **Pupils can now do Exercise 9D from Pupil Book 3.1.**

Part 3

- Say that this activity covers both multiplication and division of decimals.
- Ask pupils to give a number that makes 0.8 smaller when multiplied by that number.
- Obtain some examples and write them on the board. What is the common characteristic?
- Establish that multiplying by any value less than 1 makes 0.8 smaller.
- What about a number that makes 0.8 larger when multiplied by that number?
- Obtain some examples and write them on the board. What is the common characteristic?
- Establish that any value larger than 1 will work.
- Repeat the above procedures with 0.8 divided by a number. Establish that values greater than 1 make 0.8 smaller and values less than 1 make 0.8 larger.
- If there is time, test pupils' understanding by asking for the missing values in:
  $0.8 \times \ldots = 8$ $0.8 \div \ldots = 0.08$ $0.8 \times \ldots = 0.08$ $0.8 \div \ldots = 8$

Answers

Exercise 9D

1	**a** 0.2	**b** 0.3	**c** 0.6	**d** 0.7
	e 0.3	**f** 0.6	**g** 0.6	**h** 0.8
2	**a** 0.18	**b** 0.09	**c** 0.04	**d** 0.05
	e 0.04	**f** 0.09	**g** 0.09	**h** 0.07
3	**a** 1.21	**b** 0.65	**c** 1.61	**d** 0.95
	e 0.87	**f** 0.89	**g** 0.61	**h** 0.32
4	**a** 9.01	**b** 7.59	**c** 5.76	**d** 5.46
	e 3.06	**f** 11.02	**g** 2.36	**h** 1.28
5	**a** 8.05	**b** 1.46	**c** 3.3	**d** 1.88
	e 8.65	**f** 4.825	**g** 5.24	**h** 2.65

6 1.08 cm
7 £0.47
8 2.08 km/h
9 0.46 m
10 £1.35

Investigation: Spot the link

A	**a** 6.85	**b** 685	**c** 6.85	**d** 0.685
B	**a** 46	**b** 4.6	**c** 0.46	**d** 46

Lesson 9.5 Solving problems

Learning objective
- To solve real-life problems involving multiplication or division

Resources
- Pupil Book 3.1, pages 151–153
- Intervention Workbook 2, pages 22–23
- Homework Book 3, section 9.5
- Online homework 9.5, questions 1–10

Links to other subjects
- **Design and technology** – to calculate a range of measure for costing and producing products
- **Art** – to work with proportion problems for design purposes

Key words
- No new key words for this topic

Problem solving and reasoning help
- The questions in Exercise 9E of the Pupil Book are nearly all **PS** questions and make use of pupils' calculation skills in a combination of abstract and real-life problems. The abstract problems include multi-step problems that draw on different mathematics areas. The real-life contexts involve measures and money. Encourage pupils to discuss their methods and show their working. Pupils working at this level would benefit from regular mini exercises and/or guided group work to check progress. The investigation at the end of the exercise gives pupils the opportunity to practise their skills while exploring a magic square. Pupils may have met magic squares before, so draw on this prior learning before pupils start the investigation.

Common misconceptions and remediation
- Pupils often struggle to decode word problems to identify the mathematics they need to use. Provide plenty of opportunity for pupils to discuss word problems to identify the mathematics required independently. Encourage pupils to develop a set of transferable strategies such as marking the text to identify key words and information. Discuss how pupils can apply these strategies to a range of problems across mathematics, particularly in real life.

Probing questions
- What is the key information in this problem?
  - What information do you not need?
  - What other information might you need to work out or find the answer?
  - What mathematics will you use?
  - Is it a single-step problem or a multi-step problem?
- Is the problem easy or difficult? What makes the problem easy or difficult?

Part 1
- This activity revisits learning from earlier lessons in this chapter.
- Prepare several cards with a range of whole and decimal numbers. Choose the numbers based on the group. Do this as a class or in groups to allow further differentiation.
- Choose two cards and ask pupils to multiply or divide the numbers on them.
- Check pupils' answers. Then move on to a new problem.

Part 2

- This lesson is about problem-solving strategies.
- Work through examples 12 and 13 on page 152 of the Pupil Book. Example 12 is a best-buy problem, which are very common in exams as well as in real life. It would be useful for pupils to be able to recognise these types of problems and the mathematics needed to solve them.
- Pupils may benefit from discussion about the strategies for each question in Exercise 9E.
- **Pupil can now do Exercise 9E from Pupil Book 3.1.**

Part 3

- Revisit some of the strategies used to solve some of the problems in Exercise 9E. Working in pairs, ask pupils to think about the following points:
- Are there other types of questions (such as best-buys) with which pupils could identify?
- Ask pupils if there are key words that will help them to decide what mathematics to use.
- Can they use strategies like checking or inverse operations to help check their answers?
- Can they give some examples of how they could do this?
- Take some feedback. Try to make connections to all the lessons in this chapter.

Answers

Exercise 9E

1 any two odd numbers with a sum of 48, for example: 1 and 47 or 23 and 25
2 42
3 4
4 **a** 5 **b** 5 **c** 7 or 23
d any two numbers with a difference of 15, for example: 3 and 18 or −1 and 14
5 6 × 45p = £2.70, so the multipack is the cheaper way to buy them.
6 53 years old
7 the mass of the odd-numbered counters → 5 × 6
the total mass of the counters → 10 × 6
the mass of the counters that are not green → (5 + 2) × 6
8 60
9 £1
10 Rebecca is 12 years old and Oliver is 24.
11 1.5p

Investigation: Magic square

A

16	2	3	13
5	11	10	18
9	7	6	12
4	14	15	1

B the four centre squares
If you split the 4 by 4 square into four 2 by 2 squares along its axes of symmetry, then each of the 2 by 2 squares adds up to the magic number.
the middle 2 squares of the top row plus the middle 2 squares of the bottom row
the middle 2 squares of the left hand column plus the middle two squares of the right hand column

Review questions

(Pupil Book pages 154–155)

- The review questions will help to determine pupils' abilities with regard to the material within Chapter 9.
- These questions also draw on the mathematics covered in earlier chapters of the book to encourage pupils to make links between different topics.
- The answers are on the next page of this Teacher Pack.

Mathematical reasoning – Paper

(Pupil Book pages 156–157)

- This activity uses the context of paper, with which pupils may be very familiar. All the information pupils need is provided in the text in the Pupil Book, but it is quite complex. Pupils will need to read the questions very carefully to decide on the information that they will need and what mathematical skills to use in each case.
- The questions move freely between fractions and decimals. This is something that pupils need to be comfortable with, which also develops their conceptual understanding of fractions and decimals being ways of expressing parts of a whole.
- As a warm-up you could work with pupils to ask them to identify the key information and the mathematical skills they will need to answer the questions.
- Pupils should work in small groups to encourage discussions about the problem.
- Ask **more able** pupils to design their own multi-step problem using the information provided, or an alternative set of information about a different real-life context. You could ask pupils to draw on some of the other skills they have learnt in the lesson.

Answers to Review questions

1 **a** 680 **b** 7400 **c** 92 000 **d** 71 300
e 85.9 **f** 1.18 **g** 0.584 **h** 3.9

2 any two odd numbers that add up to 56, for example: 1 and 45 or 23 and 33

3 **a** 2.4 **b** 23.2 **c** 8.2 **d** 38.4 **e** 4.0

4 **a** 3.37 **b** 35.08 **c** 7.15 **d** 42.03 **e** 1.00

5 **a** 1.6 m **b** 1.35 m **c** 160 cm

6 The six-pack is a better value at 21.7p per tin. The four-pack costs 22.5p per tin.

7 Brushup is cheaper (£25) than Kleengo (£29).

8 **a** 18.5 **b** 25.8 **c** 23.28 **d** 36.2
e 9 **f** 30 **g** 70 **h** 80

9 **a** 10^{-3} **b** 10^{-1} **c** 10^{-2} **d** 10^{-5} **e** 10^{-6}

10 **a** 0.28 **b** 0.64 **c** 0.45 **d** 0.21 **e** 0.99

11

	Population	Land mass per person (km^2)
Vatican City	826	$53.3 \times 10^{-5} = 0.000\ 533$
Gibraltar	31 000	$21.9 \times 10^{-5} = 0.000\ 219$
Monaco	33 000	$59 \times 10^{-6} = 0.000\ 059$
Bermuda	65 000	$8.2 \times 10^{-4} = 0.000\ 82$

12 Answers may vary from those given provided each answer is accompanied by a valid reason.
a ii 90 km/h **b** ii 75° **c** ii 1.2 kg **d** ii 30 minutes

Answers to Mathematical reasoning – Paper

1 2500

2 10 cm

3

A0	A1	A2	A3	A4	A5
1188	840	594	420	297	210
840	594	420	297	210	149

4 16

5 **a** £10.80 **b** £14 **c** £14.40

6 £33.60

10 Surface area and volume of 3D shapes

Learning objectives

- How to work out the surface areas of cubes and cuboids
- How to work out the volumes of cubes and cuboids
- How to work out the volume of triangular prisms

Prior knowledge

- How to work out the areas of squares and rectangles
- That the units for area are square centimetres (cm^2) and square metres (m^2)

Context

- Remind pupils that perimeter, area and volume are used widely in many jobs and professions, from farming to astronomy. Encourage pupils to ask family and friends if they use these units of measure in their work. Pupils could also explore specific jobs on the internet. A good example is the building industry, which is totally dependent on workers being able to measure lengths and calculate areas.

Discussion point

- In what way are the perimeter and area of a shape different?
- Do you have any tips to help you to remember which is which?

Associated Collins ICT resources

- Chapter 10 interactive activities on Collins Connect online platform
- *Understanding surface area of prisms* video on Collins Connect online platform
- *Christmas cone* Wonder of Maths on Collins Connect online platform

Curriculum references

Geometry and measures

- Use the properties of faces, surfaces, edges and vertices of cubes, cuboids, prisms, cylinders, pyramids, cones and spheres to solve problems in 3D
- Derive and apply formulae to calculate and solve problems involving: perimeter and area of triangles, parallelograms, trapezia, volume of cuboids (including cubes) and other prisms (including cylinders)
- Calculate and solve problems involving: perimeter of 2D shapes (including circles), areas of circles and composite shapes

Develop fluency

- Use language and properties precisely to analyse 2D and 3D shapes

Solve problems

- Develop their mathematical knowledge, in part through solving problems and evaluating the outcomes, including multi-step problems

Fast-track for classes following a 2-year scheme of work

- Pupils should be familiar with many of the concepts in this chapter. Check pupils' understanding by giving them different examples to see if they have any problems finding the answers. Once you are happy that pupils are confident, move on to the **MR** questions towards the end of each exercise, and the investigation and problem-solving activities at the end of each exercise.

Lesson 10.1 Surface areas of cubes and cuboids

Learning objective

- To work out the surface areas of cubes and cuboids

Resources and homework

- Pupil Book 3.1, pages 159–161
- Homework Book 3, section 10.3
- Online homework 10.1, questions 1–10

Links to other subjects

- **Design and technology** – to design packaging material
- **Art** – to design 3D sculptures involving cubes and cuboids

Key words

- cube
- surface area
- cuboid

Problem solving and reasoning help

- **MS** question 8 of Exercise 10A in the Pupil Book helps pupils to extend their understanding of cubes and cuboids to real-life problems. If pupils struggle, provide concrete examples to help them visualise what they need to do. Encourage **more able** pupils to try to explain why any rules they find make sense, based on the structure of the problem.

Common misconceptions and remediation

- Pupils often confuse the concept of surface area and volume. Using concrete examples will help pupils to understand the difference.

Probing questions

- How do you find the surface area of a cuboid?
- Find three cuboids with a surface area of 48 cm^2.

Part 1

- This is a revision exercise to ensure that the pupils know the formulae for the area of different quadrilaterals. On the board or OHP, draw these quadrilaterals: a square, a rectangle, a parallelogram.
- In their books, ask pupils to write the formula for the area of each quadrilateral.
- Check pupils' answers by asking individuals to write each formula on the board.

Part 2

- Show the class a cuboid made from multi-link cubes with: length = 4 cm, width = 3 cm and height = 2 cm.
- Remind the class how to find the total surface area of the cuboid by finding the area of its six surfaces and adding them together. Draw the cuboid on the board.

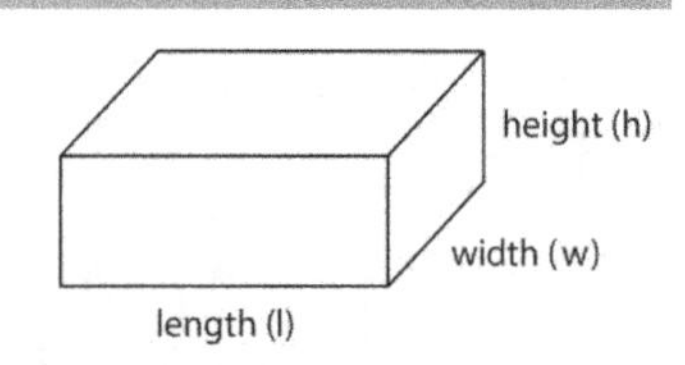

- The formula to find the total surface area of any cuboid is: $A = 2lw + 2lh + 2wh$
- Therefore, the surface area of the multi-link cuboid is:
  $A = (2 \times 4 \times 3) + (2 \times 4 \times 2) + (2 \times 3 \times 2)$
  $= 24 + 16 + 12$
  $= 52\ cm^2$
- **Pupils can now do Exercise 10A from Pupil Book 3.1.**

Part 3

- Draw a cuboid on the board:
- Ask the class to explain how to find the total surface area of the cuboid. Make sure that all pupils know how to work it out.

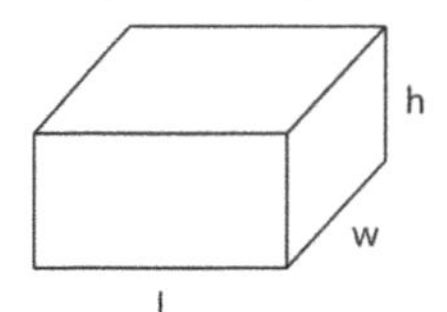

Answers

Exercise 10A

1 $(2 \times 4 \times 3) + (2 \times 4 \times 2) + (2 \times 3 \times 2) = 24 + 16 + 12 = 52\ \text{cm}^2$

2 There are 6 square faces and each one has an area of $4 \times 4 = 16\ \text{cm}^2$.
So the surface area of the cube is $6 \times 16 = 96\ \text{cm}^2$.

3 **a** $72\ \text{cm}^2$ **b** $92\ \text{cm}^2$ **c** $192\ \text{cm}^2$ **d** $46\ \text{cm}^2$

4 $94\ \text{cm}^2$

5 **a** $6\ \text{cm}^2$ **b** $24\ \text{cm}^2$ **c** $150\ \text{cm}^2$ **d** $216\ \text{cm}^2$

6 $13.5\ \text{m}^2$

7 $700\ \text{cm}^2$

8 $39\ \text{m}^2$

Investigation: An open box problem

A $2 \times 2 = 4\ \text{cm}^2$, $4 \times 4 = 16\ \text{cm}^2$, $192 - 16 = 176\ \text{cm}^2$

B

Size of squares cut out	Area of the four squares (cm^2)	Surface area of box (cm^2)
1 cm by 1 cm	$4\ \text{cm}^2$	$188\ \text{cm}^2$
2 cm by 2 cm	$16\ \text{cm}^2$	$176\ \text{cm}^2$
3 cm by 3 cm	$36\ \text{cm}^2$	$156\ \text{cm}^2$
4 cm by 4 cm	$64\ \text{cm}^2$	$128\ \text{cm}^2$
5 cm by 5 cm	$100\ \text{cm}^2$	$92\ \text{cm}^2$

Lesson 10.2 Volume formulae for cubes and cuboids

Learning objectives
- To use simple formulae to work out the volume of a cube or cuboid
- To work out the capacity of a cube or cuboid

Resources and homework
- Pupil Book 3.1, pages 162–165
- Homework Book 3, section 10.2
- Online homework 10.2, questions 1–10

Links to other subjects
- **Design and technology** – to design packaging material
- **Art** – to design 3D sculptures involving cubes and cuboids

Key words
- capacity
- litre
- height
- volume

Problem solving and reasoning help
- **MR** questions 7 and 8 in Exercise 10B of the Pupil Book require pupils to find missing lengths of cuboids, given the dimensions of two sides and/or the volume. **Less able** pupils may struggle with this, so you may need to model some examples showing pupils how to set up the formula.

Common misconceptions and remediation
- Pupils often confuse the concept of surface area and volume. Use concrete examples to help pupils understand the difference.

Probing questions
- How do you find the volume of a cuboid?
- Find three cuboids with a volume of 48 cm^2.

Part 1
- Write the number 12 on the board.
- Ask individuals to come to the board and write three numbers that have a product of 12, allowing repeats. Remind pupils that 'product' means 'multiply'.
- Examples of numbers are: $1 \times 1 \times 12$, $1 \times 2 \times 6$, $1 \times 3 \times 4$, $2 \times 2 \times 3$.
- Repeat the activity using different numbers.

Part 2
- Explain to the class that *volume* is the amount of space inside a 3D shape.
- Show the class the multi-link cube and explain that the volume is made from 24 cubes.
- A quick way to find the volume of a cuboid is to multiply its length by its width by its height.
- Using the same diagram:
  The volume of a cuboid = length × width × height
  $V = l \times w \times h = lwh$

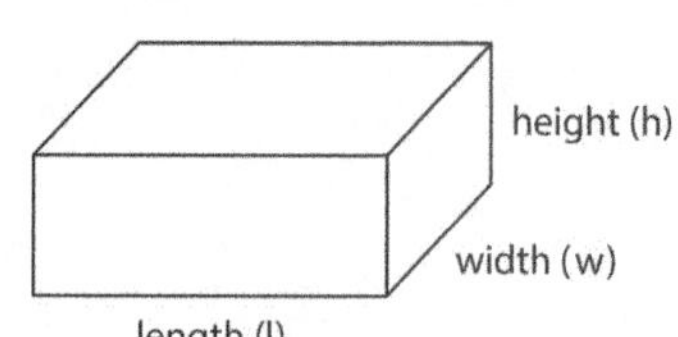

- The metric units of volume that are commonly used are: cubic millimetre (mm^3), cubic centimetre (cm^3), cubic metre (m^3).
- So, for example, the volume of the multi-link cuboid shown could be: $V = 4 \times 3 \times 2 = 24$ cm^3.
- Explain to pupils that the *capacity* of a 3D shape is the volume of liquid or gas it can hold. The metric unit of capacity is the litre (*l*). Pupils should write the following:
  100 centilitres (cl) = 1 litre, 1000 millilitres (ml) = 1 litre.

- Show the class various objects that are used for capacity, for example: a 5 ml teaspoon, a 15 ml tablespoon, a 250 ml measuring cup or a 70 cl wine bottle.
- Tell pupils that they should also note these metric conversions for capacity: 1 litre = 1000 cm^3, 1 ml = 1 cm^3, 1000 litres = 1 m^3.
- **Pupils can now do Exercise 10B from Pupil Book 3.1.**

Part 3

- Draw a cuboid on the board:
- Ask the class to explain how to find the volume of the cuboid, making sure that all pupils are confident in their explanation.

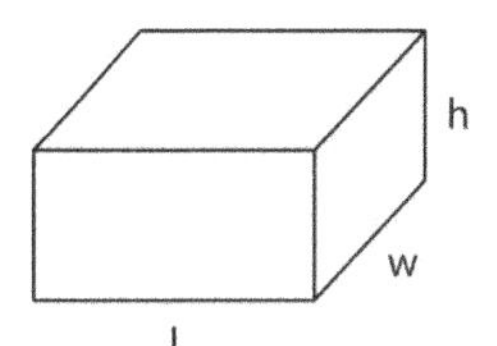

Answers

Exercise 10B

1 6 × 5 × 4 = 120 cm^3
2 5 × 5 × 5 = 125 cm^3
3 **a** 168 cm^3 **b** 360 cm^3 **c** 3 m^3
4 **a** 64 cm^3 **b** 216 cm^3 **c** 1728 cm^3
5 **a** 16 *l* **b** 30 *l* **c** 120 *l*
6 48 m^3
7 **a** 3 cm **b** 8 cm **c** 8 m
8 10 cm

Investigation: Painted cubes

A **a** 0 **b** 0 **c** 0 **d** 8
B **a** 1 **b** 6 **c** 12 **d** 8
C **a** 8 **b** 24 **c** 24 **d** 8
D

Size of yellow cube	Number of cubes	No faces painted red	One face painted red	Two faces painted red	Three faces painted red
2 by 2 by 2	8	0	0	0	8
3 by 3 by 3	27	1	6	12	8
4 by 4 by 4	64	8	24	24	8

Lesson 10.3 Volumes of triangular prisms

Learning objective

- To work out the volume of a triangular prism

Resources and homework

- Pupil Book 3.1, pages 166–169
- Homework Book 3, section 10.2
- Online homework 10.3, questions 1–10

Links to other subjects

- **Design and technology** – to calculate the volumes of prism-shaped packaging materials

Key words

- triangular prism

Problem solving and reasoning help

- Pupils will need to apply their learning from this lesson to real-life problems. Help **less able** pupils to pick out the key words in the questions, which should make it easier for them to be able to work out the problems. Make sure pupils know how to calculate the area of a triangle.

Common misconceptions and remediation

- Pupils sometimes confuse area with volume. Pupils occasionally forget to check that all the units are the same before calculating. Remind pupils to provide units with their answers.

Probing questions

- Explain how you would calculate the volume of a triangular prism.
- If we are given the height and volume of a prism, what else can we work out?

Part 1

- Show the class various 3D shapes such as a: cube, cuboid, square-based pyramid, tetrahedron, triangular prism, cylinder, sphere.
- Ask pupils to identify each shape and to spell each name. Write all the names on the board.

Part 2

- Tell pupils that they will now learn how to find the volume of a triangular prism. Refer to the diagram and show pupils other prism shapes.

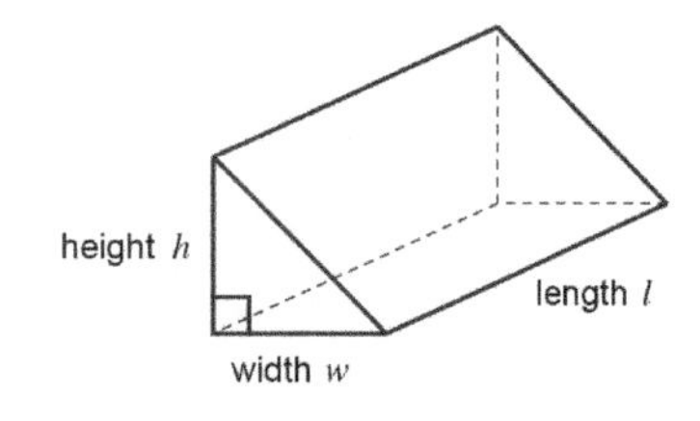

- Ask pupils to explain what is similar about the shapes.
- Ask the class to write the definition of a prism into their books: A prism is a 3D shape, which has exactly the same 2D shape running all the way through it.
- Point out to pupils that if they cut through a prism anywhere at right angles to its length, they will see a 2D-shape, which is known as the *cross-section*.
- Ask for examples of triangular prisms (packets, triangular chocolate slab, door wedge).
- Remind the class that a cuboid is an example of a prism. If possible, show the class models of different prism shapes.

- Draw the diagram of the prism on the board. The volume is given by:

  $V = \frac{lwh}{2}$

  $= \frac{15 \times 6 \times 8}{2}$

  $= \frac{720}{2}$

  $= 360 \text{ cm}^3$

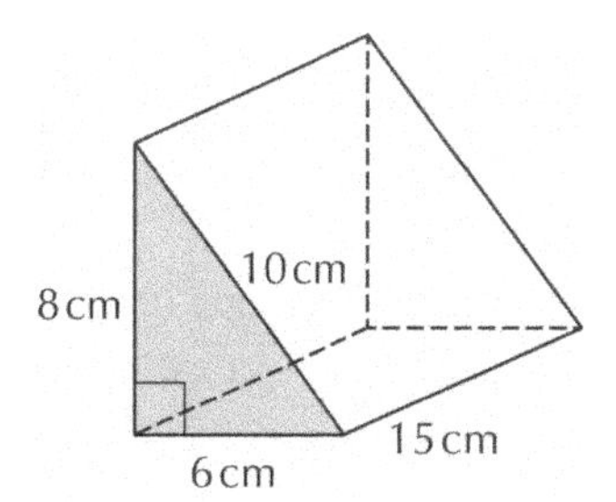

- **Pupils can now do Exercise 10C from Pupil Book 3.1.**

Part 3

- Ask pupils to give the definition of a prism. Then invite pupils to give some everyday examples of where they might see triangular prisms.
- Ask individual pupils to explain how to find the volume of a triangular prism.

Answers

Exercise 10B

1 $\frac{6 \times 7 \times 4}{2} = \frac{168}{2} = 84 \text{ cm}^3$

2 $\frac{16 \times 12 \times 9}{2} = \frac{1728}{2} = 864 \text{ cm}^3$

3 **a** 90 cm^3 **b** 100 cm^3 **c** 288 cm^3

4 800 cm^3

5 **a** 0.3 m^3 **b** 0.72 tonnes

6 6 cm

Problem solving: Surface area of triangular prisms

A 36 cm^2

B 360 cm^2

Review questions (Pupil Book pages 170–171)

- The review questions will help to determine pupils' abilities with regard to the material within Chapter 10.
- These questions also draw on the mathematics covered in earlier chapters of the book to encourage pupils to make links between different topics.
- The answers are on the next page of this Teacher Pack.

Investigation – A cube investigation (Pupil Book pages 172–173)

- Pupils apply their understanding of area to a more complex problem.
- Pupils need to work methodically and be able to explain their solutions. This is a good transferable skills objective to share with pupils when doing this investigation. Ask pupils to share not only their solutions but also *how* they approached working on the problem.
- Pupils should work in small groups to encourage discussions about the problem. Make sure that pupils have a collection of cubes and some centimetre isometric dotted paper.
- The cubes, in particular, will help **less able** pupils to visualise the problems.
- You could use strategies such as reading images to structure pupils' responses to the final questions. For example:
    - What is the least surface area for the different shapes you have made? Explain why.
    - What do you think the surface areas of a 3D shape made from five cubes could be?
    - Work out the least surface area of a 3D shape made from six cubes.
- Pupils should write these questions in the centre of a large sheet of paper. Around the outside, give fixed sub-questions to which they should respond, for example, 'What do you know? What does this mean? Can you see any patterns? Can you explain these patterns?' This will scaffold pupils' thinking about the problem.
- You could ask **more able** pupils to provide more detailed justifications of any rules, by revisiting the structure of the problem.

Answers to Review questions

1 **a** $(2 \times 6 \times 4) + (2 \times 6 \times 3) + (2 \times 4 \times 3) = 48 + 36 + 24 = 108\ \text{cm}^2$

b $6 \times 4 \times 3 = 72\ \text{cm}^3$

2 **a i** 174 cm^2 **ii** 135 cm^3

b i 3 m^3 **ii** 210 m^3

c i 48 cm^2 **ii** 20 cm^3

3 1.5 litres

4 $\frac{10 \times 9 \times 8}{2} = \frac{720}{2} = 360\ \text{cm}^3$

5 1.8 m^3

6 **a** 5 cm **b** A = 112 cm^2 and B = 94 cm^2, so cuboid A

Answers to Investigation – A cube investigation

1

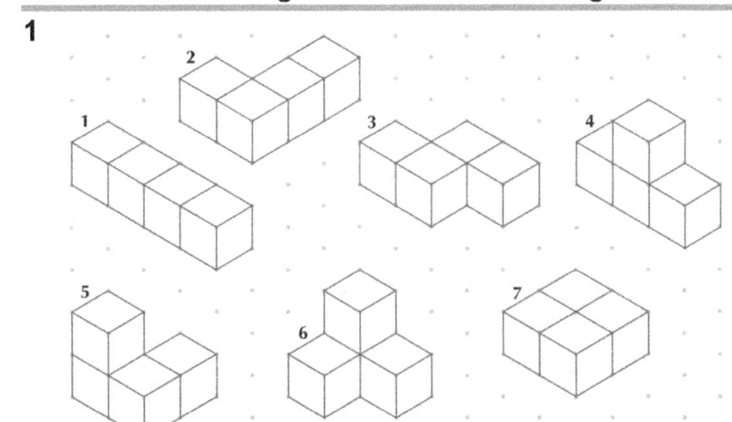

2

3D shape	1	2	3	4	5	6	7
Surface area	18 cm^2	18 cm^2	18 cm^2	18 cm^2	18 cm^2	18 cm^2	16 cm^2

3 Shape 7 has the least surface area and the rest have the same surface area. More surfaces are in contact with each other in shape 7.

4 22 cm^2 and 20 cm^2

11 Solving equations graphically

Learning objectives

- How to solve linear equations graphically
- How to use straight-line graphs to solve problems
- How to solve simple quadratic equations
- How to use quadratic graphs to solve problems

Prior knowledge

- How to draw linear graphs of the form $y = mx + c$
- How to draw a simple quadratic graph

Context

- This chapter provides examples of the fact that many equations can arise from real-life situations, and it builds on straight-line graphs and simple quadratics. Pupils are introduced to the idea that many equations that are used to model real life are difficult to solve by algebraic methods, but they are solved more easily by drawing a graph. You could show video clips such as the one at this link, of a horse jumping:
http://www.youtube.com/watch?v=mgVi78diBwQ

Discussion points

- Given the coordinates of three points on a straight line parallel to the y-axis, find the equation of the line.
- Given the coordinates of three points on a straight line such as $y = 2x$, find three more points in a given quadrant.
- If you wanted to plot the graph $y = 2x$, how would you start?

Associated Collins ICT resources

- Chapter 11 interactive activities on Collins Connect online platform
- *Ice sculptures* Wonder of Maths on Collins Connect online platform

Curriculum references

Develop fluency

- Develop algebraic and graphical fluency, including understanding linear and simple quadratic functions

Algebra

- Reduce a given linear equation in two variables to the standard form $y = mx + c$; calculate and interpret gradients and intercepts of graphs of such linear equations numerically, graphically and algebraically
- Use linear and quadratic graphs to estimate values of y for given values of x and vice versa

Fast-track for classes following a 2-year scheme of work

- Pupils may be familiar with the material in the first two lessons of this chapter. Check pupils' understanding by giving them some well-targeted questions about $y = mx + c$. If they are confident, you may want to combine Lesson 11.1 and Lesson 11.2 using the **MR** and **PS** questions, and the end of lesson activities.

Lesson 11.1 Graphs from equations in the form $y = mx + c$

Learning objectives
- To draw a linear graph from any linear equation
- To solve a linear equation from a graph

Resources
- Pupil Book 3.1, pages 175–177
- Online homework 11.1–2, questions 1–10

Links to other subjects
- **Science** – to model physical situations

Key words
- linear equation

Problem solving and reasoning help
- The investigation at the end of Exercise 11A in the Pupil Book introduces $x + y = c$ and negative gradients.

Common misconceptions and remediation
- Pupils often learn general rules without really understanding them. As a result, they may struggle to see the connection to different formats for the equation of a straight line.
- Give pupils plenty of opportunity to write equations in as many ways as they can.

Probing questions
- Can you give me the equations of graphs that pass through (0, 1)? What about [other points]?
- How would you find coordinates for this straight-line graph that are in this quadrant?

Part 1
- Ask pupils, mentally, to solve $x - 1 = 0$. Quite quickly, pupils should say: $x = 1$ (pupils could write their answers on mini whiteboards).
- Repeat with: $x + 4 = 0$ (–4), $x - 3 = 0$ (3), $x + 3 = 0$ (–3).
- Ask pupils for examples that solve $y + x = 4$.
- Encourage pupils to use negative numbers and to say that in this case there are in fact an infinite number of solutions.

Part 2
- Use the example $y = 3x + 1$ in Example 1 on page 175 of the Pupil Book. By the end of Exercise 11A pupils should have consolidated their understanding of the role of m as the gradient and c as the intercept.
- **Pupils can now do Exercise 11A from Pupil Book 3.1.**

Part 3
- Ask pupils how they know that the point (4, 6) is not on the line $y = x + 3$?
- Working in pairs, ask pupils to try and write a similar question.
- Share some examples with the class.

Exercise 11A

1 a

x	0	1	2	3
$y = x + 1$	1	2	3	4
$y = x + 2$	2	3	4	5
$y = x + 3$	3	4	5	6
$y = x + 4$	4	5	6	7

b–c **e**

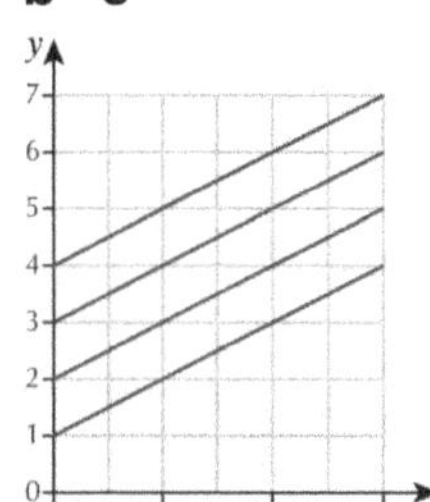

d parallel and cut the y-axis at the value added to x

2 a

x	0	1	2	3
$y = 2x + 1$	1	3	5	7
$y = 2x + 2$	2	4	6	8
$y = 2x + 3$	3	5	7	9
$y = 2x + 4$	4	6	8	10

b–c **e**

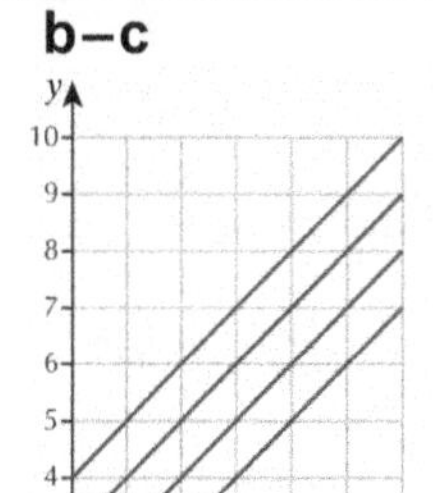

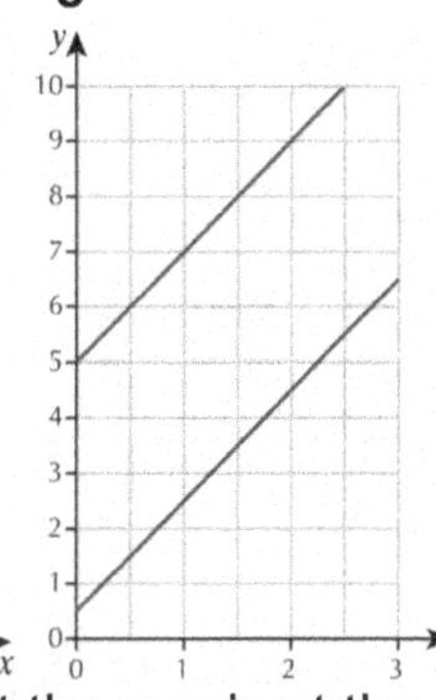

d parallel and cut the y-axis at the value added to $2x$

3 a

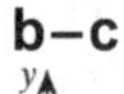

x	0	1	2	3
$y = x$	0	1	2	3
	0	2	4	6
	0	3	6	9
	0	4	8	12

b–c **e**

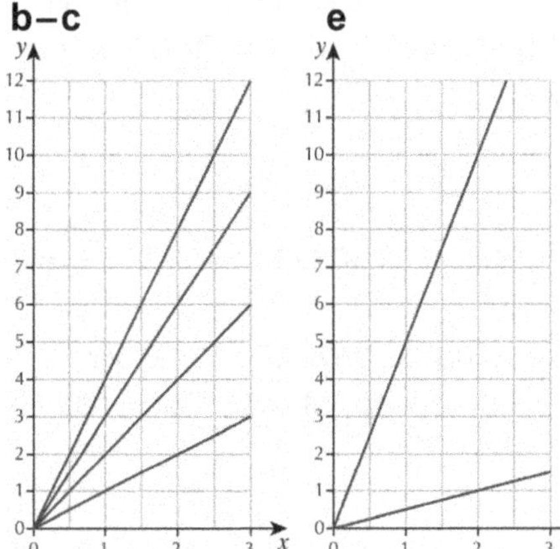

d They all pass through the origin and they get steeper as the value in front of x increases.

4 a

x	0	1	2	3
$y = x + 4$	4	5	6	7
$y = 2x + 4$	4	6	8	10
$y = 3x + 4$	4	7	10	13
$y = 4x + 4$	4	8	12	16

b–c **f**

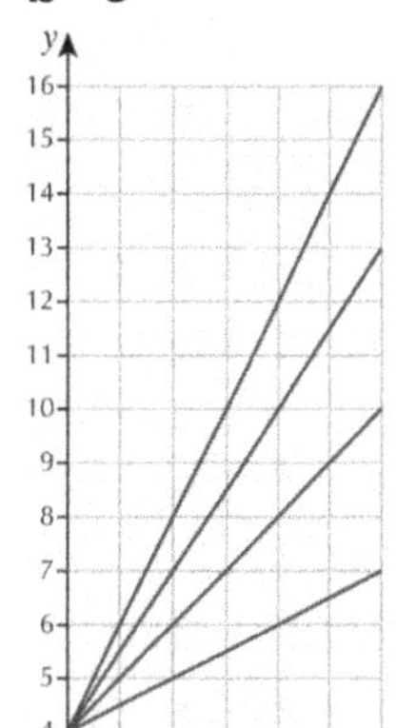

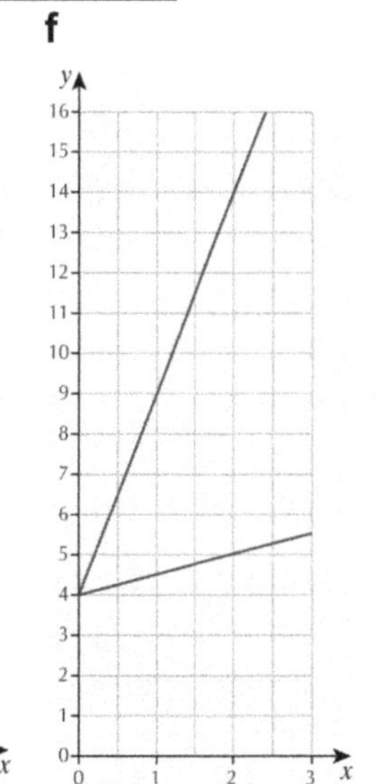

d They all cut the y-axis at (0, 4) and they get steeper as the value in front of x increases.

e how steep it is – the gradient

Investigation: The graph of $x + y = c$

A–B

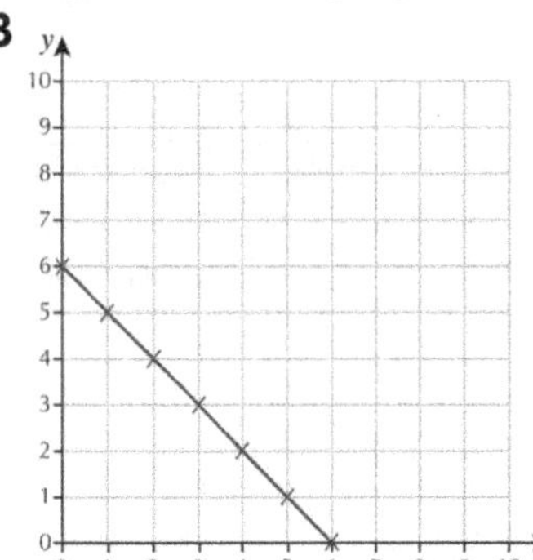

C (0, 5), (1, 4), (2, 3), (3, 2), (4, 1), (5, 0)

D

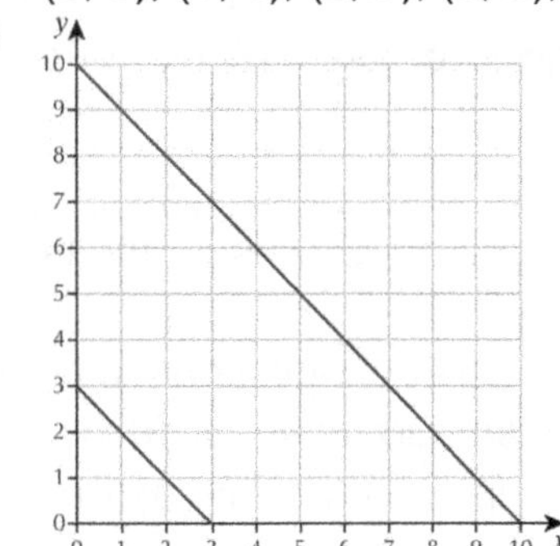

Lesson 11.2 Problems involving straight-line graphs

Learning objective
- To draw graphs to solve some problems

Resources
- Pupil Book 3.1, pages 178–180
- Online homework 11.1–2, questions 1–10

Links to other subjects
- **Science** – to find best-fit equations to model physical phenomena
- **Physical education** – to analyse flight paths to improve performance

Key words
- No new key words for this topic

Problem solving and reasoning help

- The challenge activity at the end of Exercise 11B in the Pupil Book gives pupils the opportunity to apply what they know to a more complex real-life situation. You may want to approach this as part of a class discussion, or as a guided group discussion, depending on the group. Give pupils plenty of opportunity to discuss how they make links between the words and the mathematics they have been learning.

Common misconceptions and remediation

- Pupils sometimes struggle to understand that we use mathematics to model the real world but that these models will never be exact. Provide plenty of opportunity for pupils to discuss this for a range of contexts.

Probing questions

- What can you tell about its graph from looking at its function?
- How would you find the y value for a given x value? And an x value for a given y value?
- If I wanted to plot the graph $y = 2x$ how should I start?
- How do you know that the point (3, 6) is not on the line $y = x + 2$?

Part 1

- Put up some examples of straight-line graphs and their equations in the form $y = mx + c$.
- Working in pairs, ask pupils to match the graphs to their equations.
- You could have an 'odd one out' and challenge pupils to see who can identify it fastest.

Part 2

- Tell pupils that you can often model real-life situations using straight-line graphs.
- Use Example 2 on page 178 of the Pupil Book to demonstrate this.
- This also introduces the idea of a conversion graph. Ask pupils if they can think of some other examples of when they could use graphs to convert from one type of unit to another.
- **Pupils can now do Exercise 11B from Pupil Book 3.1.**

Part 3

- Put a range of straight-line graphs on the board. Then ask pairs of pupils to write their own stories for each graph.
- Pupils could present their suggestions to the class. The class could provide formative feedback based on an agreed set of success criteria such as: appropriateness or creativity of the example, accuracy and clarity of the explanation.

- Differentiate based on the graphs used. For example, include negative gradients and intercepts. You could include examples where the gradient changes, such as those that are often used for distance–time graphs.

Answers

Exercise 11B

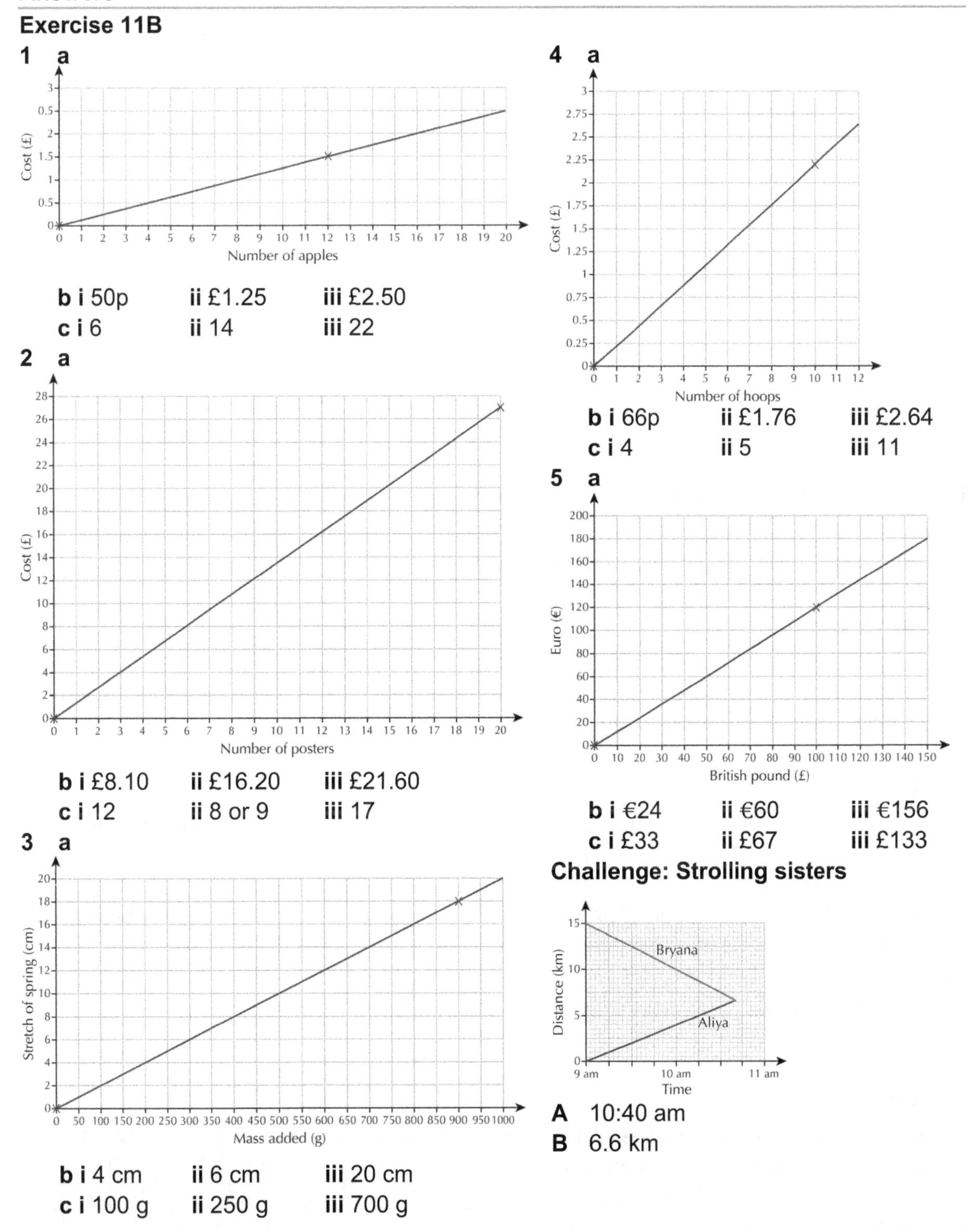

1 a

b i 50p **ii** £1.25 **iii** £2.50

c i 6 **ii** 14 **iii** 22

2 a

b i £8.10 **ii** £16.20 **iii** £21.60

c i 12 **ii** 8 or 9 **iii** 17

3 a

b i 4 cm **ii** 6 cm **iii** 20 cm

c i 100 g **ii** 250 g **iii** 700 g

4 a

b i 66p **ii** £1.76 **iii** £2.64

c i 4 **ii** 5 **iii** 11

5 a

b i €24 **ii** €60 **iii** €156

c i £33 **ii** £67 **iii** £133

Challenge: Strolling sisters

A 10:40 am

B 6.6 km

Lesson 11.3 Solving simple quadratic equations by drawing graphs

Learning objective

- To solve a simple quadratic equation by drawing a graph

Resources

- Pupil Book 3.1, pages 181–184
- Homework Book 3, section 11.3
- Online homework 11.3–4, questions 1–10

Links to other subjects

- **Science** – to find approximate solutions to complex problems to help to predict the outcomes of complex physical phenomena
- **Physical education** – to analyse flight paths to improve performance

Key words

- No new key words for this topic

Problem solving and reasoning help

- The investigation at the end of Exercise 11C in the Pupil Book considers the three possible outcomes when solving quadratic equations. This is built on in Part 3 of this lesson.

Common misconceptions and remediation

- Pupils may struggle to accept that a quadratic equation may have no solutions. Use the graphs to help pupils visualise why this is the case.

Probing questions

- Why would this quadratic have no, one, or two solutions? How do you know from the graph?
- Convince me that there are no coordinates on the graph of $y = 3x^2 + 4$ that lie below the x-axis.
- How can you identify a quadratic function from its graph?

Part 1

- Give pupils the factors of 12 in pairs and the factors of 15 on each side of a vertical line:

1	2	3	|	1	3
12	6	4	|	15	5

- Ask pupils to find a combination of products using one pair of factors from each side that equal 28. You may need to demonstrate the first example: $2 \times 5 + 6 \times 3 = (10 + 18 = 28)$
- Repeat with: 29 $(3 \times 3 + 4 \times 5)$; 27 $(3 \times 5 + 4 \times 3$ or $1 \times 15 + 12 \times 1)$; 41 $(1 \times 5 + 12 \times 3)$.
- Now ask for the total 8. Suggest using negative values: $2 \times -5 + 6 \times 3 = 8$
- Repeat with: 24 $(2 \times 15 + 6 \times -1)$; –3 $(3 \times -5 + 4 \times 3)$. Do more examples if there is time.

Part 2

- Tell pupils that you will use a table to help plot the graph of the equation: $y = x^2 + 2x - 1$.
- Say that this is similar to the approach they use when plotting the graphs of straight-line functions but that with quadratics it helps to calculate all the individual elements separately and then combine them. Display the table in Example 3 on page 181 of the Pupil Book. Describe how to substitute for each value of x into x^2 and $3x$ to determine the y value.

x	−4	−3	−2	−1	0	1	2
x^2	16	9	4	1	0	1	4
$2x$	−8	−6	−4	−2	0	2	3
−1	−1	−1	−1	−1	−1	−1	−1
$y = x^2 + 2x - 1$	7	2	−1	−2	−1	2	6

- Complete the values as a class or let pupils do this individually.
- Now take the pairs of (x, y) coordinates from the table, plot each point on a grid, and join the points.
- **Pupils can now do Exercise 11C from Pupil Book 3.1.**

Part 3

- Write $(x+2)(x+3) = 0$ on the board, together with the graph $y = x^2 + 5x + 6$.
- Ask pairs to discuss how they know that the graph and the equation are related.
- Pupils should be able to identify that it crosses the x-axis at $x = -2$ and $x = -3$
- If necessary, lead the discussion as a class or guided group work discussion.
- Using a dynamic software package such as *Geogebra* will help with this explanation: **http://www.geogebra.org/cms/en/**. Pupils could also use the package to explore the graphs of other quadratics.

Answers

Exercise 11C

1 **a** Check pupils' graphs. **b** 1.25 **c** 0.7 and −2.7
d $y = -1$ **e** 0.4 and −2.4

2 **a** Check pupils' graphs. **b** −1.25 **c** 0.8 and −3.8
d $y = -2.25$ **e** 0.3 and −3.3

3 **a** Check pupils' graphs. **b** 3.75 **c** 1.6 and −0.6
d $y = -0.25$ **e** 2.8 and −1.8

4 **a** Check pupils' graphs. **b** −0.75 **c** 2.7 and −0.7
d $y = -1$ **e** 3.6 and −1.6

5 **a** Check pupils' graphs. **b** −1.25 **c** 3.6 and −0.6
d $y = -2.25$ **e** 3.8 and −0.8

Investigation: You can't solve them all

A

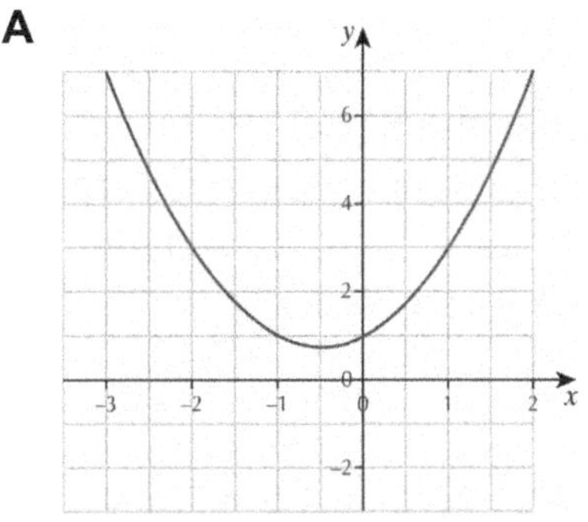

B **a** 1.6 and −2.6 **b** 0.6 and −1.6

C The graph does not cut the x-axis.

D pupils' own work

Lesson 11.4 Problems involving quadratic graphs

Learning objective

- To solve problems that use quadratic graphs

Resources

- Pupil Book 3.1, pages 185–187
- Online homework 11.3–4, questions 1–10

Links to other subjects

- **Physical education** – to analyse the paths of projectiles

Key words

- No new key words for this topic

Problem solving and reasoning help

- The investigation at the end of Exercise 11D in the Pupil Book gives pupils the opportunity to apply their learning to some more complex examples involving quadratics. The investigation looks at a very familiar situation. However, it is one that they might not have considered in this way before. Ask pupils how this information could be used.

Common misconceptions and remediation

- Pupils often try to learn rules without understanding them. As a result, pupils may find it difficult to transfer their understanding to more complex examples, including real-life applications.

Probing questions

- Why does a quadratic graph have line symmetry?
- Show me an example of a quadratic graph that:
  - has line symmetry about the y-axis
  - does not have line symmetry about the y-axis.

Part 1

- Write a range of linear equations (simple and complex multi-step) on the board.
- Ask pairs of pupils to pick some examples to solve. You may want to direct some pupils for support or challenge. Say that you want explanations that focus on how pupils solved the equations, with clear justification of why pupils tackled the equations in a particular way.

Part 2

- Tell pupils that many problems can be solved from graphs that are shaped like quadratic curves. You could use the example of a horse jumping from the opening page of this chapter: **http://www.youtube.com/watch?v=mgVi78diBwQ**. Or use a similar example from a football or tennis match.
- Work through Example 4 on page 185 of the Pupil Book to demonstrate this.
- Give pupils plenty of opportunity to ask questions as you model the example.

Part 3

- Put some examples of quadratics on the board with no, one and two solutions.
- Ask pupils to discuss what they notice about the graphs.
- Ask them which examples have a solution to $y = 0$. Do any have more than one solution?

Answers

Exercise 11D

1 **a**

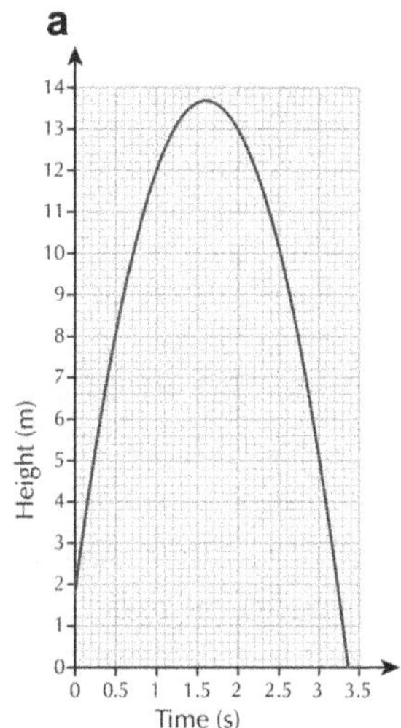

b 13.5 m

c 1.5 seconds

d 3.5 seconds

2 **a**

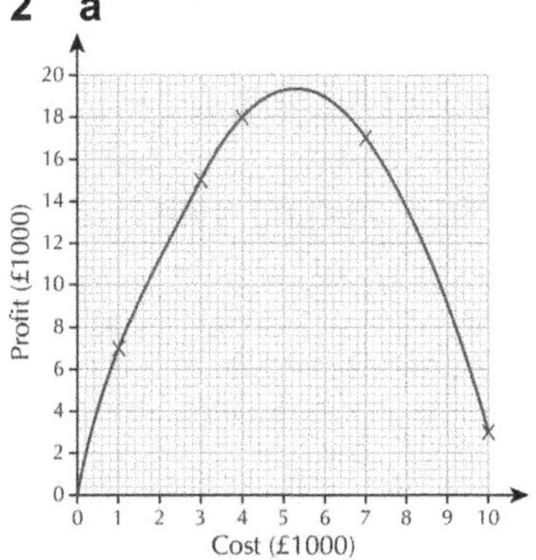

b around £19 000

c around £5200

d about £1600 or £8600

3 **a**

b approx. £2350

c about 7 m

d about 2.2 m and 13 m

4 **a**

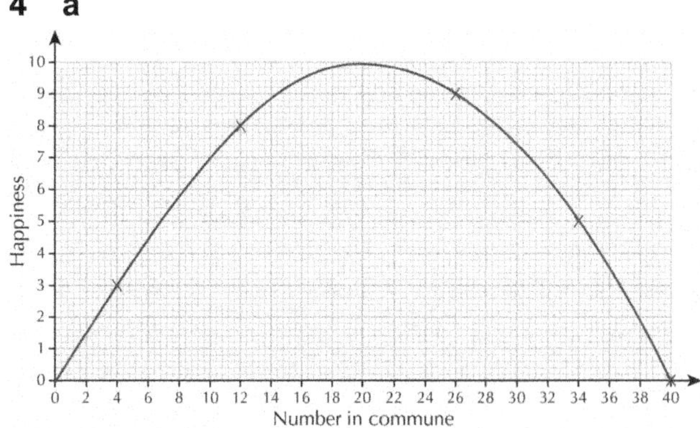

b 20

c 9

d between 8 and 32

Investigation: Cooling pies

A by drawing a graph of the data in the table and joining them with a smooth curve; then draw a line from the time up to the curve, then across to the temperature axis

a 30 °C **b** 23 °C **c** 19 °C

B between $2\frac{1}{2}$ minutes and 4 minutes after being taken out of the oven

Review questions (Pupil Book pages 188–189)

- The review questions will help to determine pupils' abilities with regard to the material within Chapter 11.
- These questions also draw on the mathematics covered in earlier chapters of the book to encourage pupils to make links between different topics.
- The answers are on the next page of this Teacher Pack.

Problem solving – Squirrels (Pupil Book pages 190–191)

- Pupils often ask why they have to do mathematics that is not familiar to them. Say that using graphs to monitor wildlife is a good example of how mathematics can be used to benefit society.
- All the information pupils need is provided in the Pupil Book text. However, it is quite dense. How much support you give pupils with deciphering the information will depend on the group. You may want to differentiate this by using guided group work to support or challenge pupils.
- As a warm-up to this activity, ask some questions about how pupils think graphs might be used in real life. You could use the clip at the Gapminder website: **http://www.gapminder.org/videos/200-years-that-changed-the-world-bbc/**
- Pupils could extend this activity by using the internet to explore a different species of animal or a different context such as population change.
- Pupils could then present what they know to the groups using graphs as part of the presentation.
- Pupils working at this level are likely to need some initial direction before starting their independent research.

Answers to Review questions

1 a

x	0	1	2	3
$y = 3x + 1$	1	4	7	10
$y = 3x + 2$	2	5	8	11
$y = 3x + 3$	3	6	9	12
$y = 3x + 4$	4	7	10	13

b–c

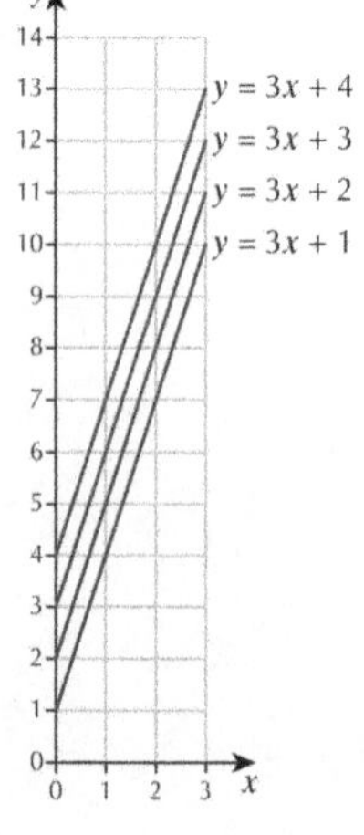

d They are parallel and cut the y-axis at the value added to the $3x$.

e

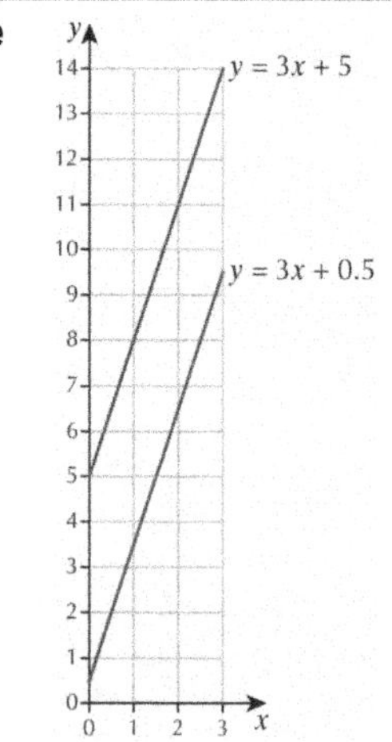

2 **a** $y = 29$ **b** $y = -11$
c $y = x^2$ and $y = x^2 - 4$ because 5^2 and $(-5)^2$ are both equal to 25.

3 **a** $A = lb = 8 \times 2 = 16$ cm²

b for example, rectangles of 1 cm × 9 cm, 3 cm × 7 cm, 4 cm × 6 cm

c The rectangles above have areas of 9 cm², 21 cm², 24 cm².

d pupils' own values, for example:

Longest side (cm)	8	9	7	6
Area (cm²)	16	9	21	24

e

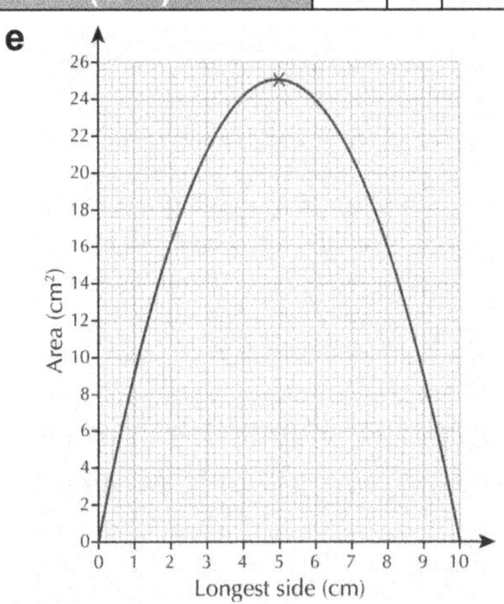

f 25 cm² **g** 5 cm

4 **a** Check pupils' graphs.

b −1.75 **c** 4.4 and −0.4

d $y = -4$ **e** 4.2 and −0.2

5 **a**

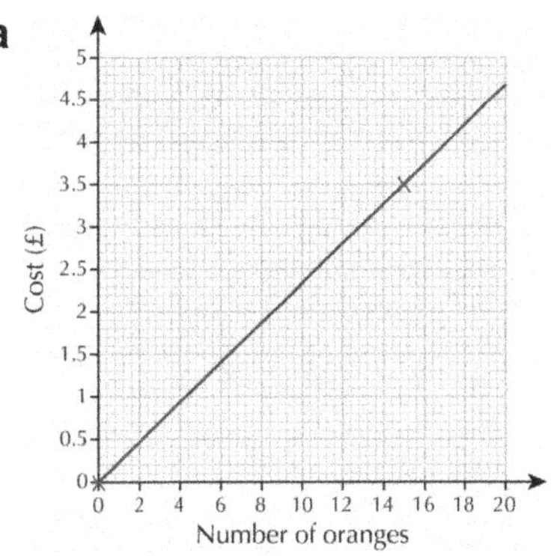

b i 93p **ii** £2.10 **iii** £4.67

c i 4 **ii** 6 **iii** 19

6 **a**

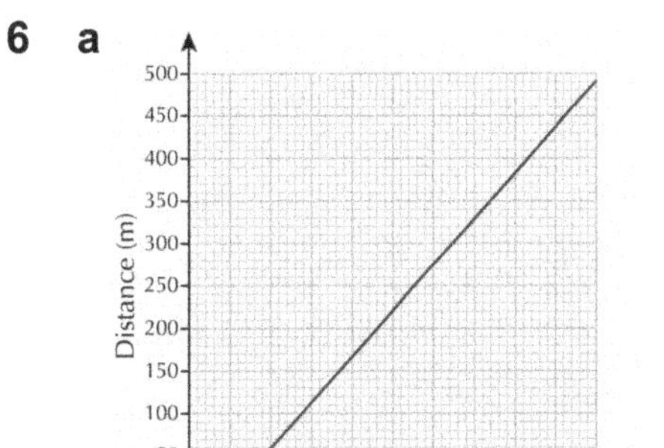

b about 8 seconds

c about 78 m

7 6.2 cm

Answers to Problem solving – Squirrels

1 **a**

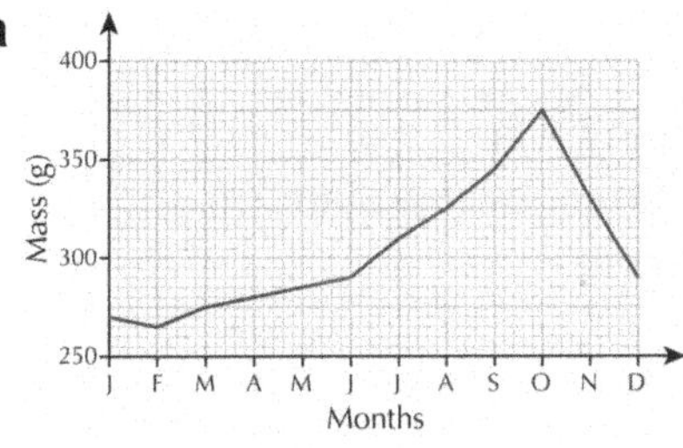

b June and July

c January/February and May

d preparing for winter hibernation

e for example:

- grey squirrels are heavier all year round
- red squirrels start to increase in mass earlier in the year
- grey squirrels increase their mass by a bigger proportion than red squirrels

2 **a** around 185–190 mm

b positive correlation

c i

ii around 260 mm

iii This is outside the range of the data supplied.

12 Distance, speed and time

Learning objectives

- How to solve problems involving distance, speed and time

Prior knowledge

- That distance is how far you travel
- That speed is how fast you are travelling
- That time is how long you spend travelling
- How to draw graphs on a grid

Context

- This chapter teaches pupils how to calculate with different measures. Pupils are introduced to the relationship between speed, distance and time. Pupils learn how to solve problems involving these measures.

Discussion points

- What does kmph or km/h mean?
- What does mps or m/s mean?
- What is mass? Is mass the same as weight?
- Would two different objects that have the same mass always have the same volume?

Associated Collins ICT resources

- Chapter 12 interactive activities on Collins Connect online platform
- *Drawing and interpreting travel graphs* video on Collins Connect online platform
- *Windmills* Wonder of Maths on Collins Connect online platform

Curriculum references

Develop fluency

- Select and use appropriate calculation strategies to solve increasingly complex problems

Reason mathematically

- Extend and formalise their knowledge of ratio and proportion in working with measures and geometry, and in formulating proportional relations algebraically

Number

- Use standard units of mass, length, time, money and other measures, including with decimal quantities

Ratio, proportion and rates of change

- Change freely between related standard units (for example, time, length, area, volume/capacity, mass)
- Use compound units such as speed, unit pricing and density to solve problems

Fast-track for classes following a 2-year scheme of work

- The material in all three lessons of this chapter will be new to pupils. However, if you feel that your pupils are confident with the different units for speed, distance and time, then you could combine the three lessons into one lesson. You could do this by using the speed, distance, time triangle.

Lesson 12.1 Distance

Learning objectives

- To work out the distance travelled in a certain time at a given speed
- To use and interpret distance–time graphs

Resources

- Pupil Book 3.1, pages 193–196
- Intervention Workbook 3, pages 73–76
- Online homework 12.1, questions 1–10

Links to other subjects

- **Science** – to calculate questions involving distance
- **Physical education** – to make calculations involving distance in athletic events

Key words

- average speed
- distance–time graph
- speed
- distance
- km/h
- time

Problem solving and reasoning help

- **PS** question 9 of Exercise 12A in the Pupil Book requires pupils to read and interpret a distance–time graph. Model some examples for **less able** pupils, discussing the key elements such as the units on each axis, what the gradient of the line represents and how to use it to read the time and distance.

Common misconceptions and remediation

- Pupils often make mistakes with the units in questions about distance. Remind pupils that the clue for the distance is given in the units of speed.

Probing questions

- Write some questions that are easy to solve, and some questions that are difficult to solve.
- How can you use information given in a problem to help you work out the units for distance?

Part 1

- Ask: 'Who knows how many miles is equivalent to 8 km?' (5 miles)
- Use this fact to ask quick-fire questions about equivalences of the following:

Kilometres	16	24	32	40	64	80	96
Miles	10	15	20	25	40	50	60

- Discuss how pupils need to think: 'How many 8s? Then multiply that by 5.'
- Ask for approximations such as 10 km and 60 km. Look for approximations of 8s. So, 10 km is just over one 8, which makes it just over 5 miles. So, we can call it 6 miles. For 60 km, divide by 8, giving $7\frac{1}{2}$, which gives $5 \times 7\frac{1}{2} = 35 + 2\frac{1}{2} = 37\frac{1}{2}$. Round to 38 miles.
- Finish by using approximation to convert the following mentally:

Kilometres	20	35	50	70	90	100	200
Miles	13	22	31	44	56	63	125

- The concern is to find approximations, so any answer 'close to the correct mileage' is acceptable. The intention is to practise mental division by 8 and mental multiplication by 5.

Part 2

- Introduce pupils to the formulae for solving problems involving speed, distance and time using the triangle notation. Discuss how pupils can use the formula to calculate distance, given speed and time:

distance = speed × time

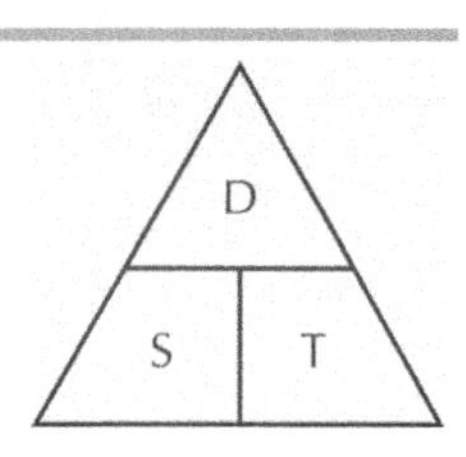

- Work through this problem with pupils: Bradley cycles at an average speed of 26 mph. If he cycles for 5 hours 30 minutes, how far does he travel? (Distance = speed multiplied by time, so: 26 mph × 5.5 hours = 143 miles)

- Work through examples 1 and 2 on pages 193 and 194 of the Pupil Book.
- Now draw a pair of axes labelled 'Time' (horizontal axis), and 'Distance' (vertical axis).
- Ask: 'What might the graph look like if it represented a car being driven at a steady speed?' (Straight line.) If you want to discuss that a steeper line represents a faster speed, create values for the graph in order to show this.
- Draw a similar pair of axes and ask: 'What shape would the graph have if the car were slowing down?' (A curve that starts with a positive gradient; gradually becomes horizontal)
- Draw another pair of axes and ask what shape the graph would be for a car accelerating from standstill to a steady speed. (A different type of curve with a horizontal gradient that gradually becomes more positive, then stays constant (a straight line) at the steady speed.)
- Discuss with the class that the graphs illustrate typical speeds, but that in real life there would be a lot of changes in the speed of the car resulting in different graphs to those drawn.
- **Pupils can now do Exercise 12A from Pupil Book 3.1.**

Part 3

- Draw a pair of axes labelled 'Time' (horizontal axis), and 'Distance' (vertical axis).
- On the axes, ask a pupil to draw a graph representing their journey to school that day.
- Discuss this graph and if it actually shows, for example, variances in speed, stationary times.
- If there is time, ask a pupil to sketch a graph of a plane journey from London to Amsterdam.

Answers

Exercise 12A

1 **a** $d = 120 \times 2 = 240$ km
b $d = 120 \times 4 = 480$ km
c $d = 120 \times 5\frac{1}{2} = 660$ km

2 **a** 80 km **b** 105 km **c** 30 km **d** 32 km

3 36 km

4 1920 km

5 20 km

6

	Speed	Time	Time as a decimal	Distance travelled
a	40 km/h	$1\frac{1}{2}$ hours	1.5 hours	60 km
b	60 km/h	$2\frac{1}{4}$ hours	2.25 hours	135 km
c	100 km/h	$\frac{3}{4}$ hour	0.75 hours	75 km
d	80 km/h	2 hours 15 minutes	2.25 hours	180 km
e	10 km/h	30 minutes	0.5 hours	5 km
f	4 km/h	3 hours 45 minutes	3.75 hours	15 km

7 **a** speed = 40 km/h
time = 30 minutes
$d = s \times t$
$= 40 \times 0.5$
$= 20$ km

b speed = 60 km/h
time = 1 hour 30
$d = s \times t$
$= 60 \times 1.5$
$= 90$ km

8 **a**

Time (h)	0	1	2	3	4
Distance (km)	0	20	40	60	80

b

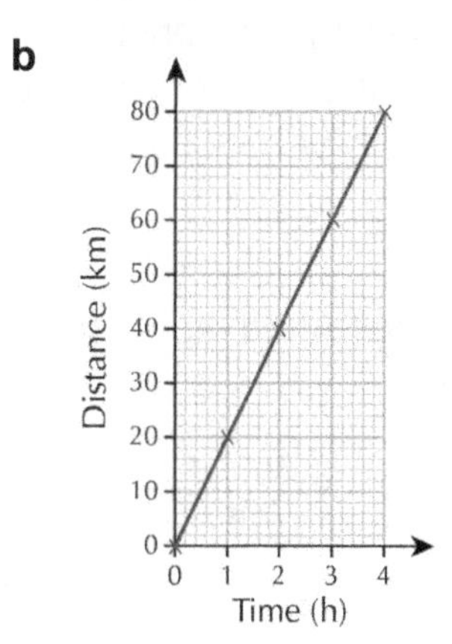

9 **a** 20 km/h
b The motorcyclist has stopped.
c 25 km/h

Challenge: Changing kilometres into miles

A **a** 15 miles **b** 50 miles
c $62\frac{1}{2}$ miles **d** $112\frac{1}{2}$ miles

B 60 miles

Lesson 12.2 Speed

Learning objective

- To work out the speed of an object, given the distance travelled and the time taken

Resources

- Pupil Book 3.1, pages 196–198
- Intervention Workbook 3, pages 73–76
- Homework Book 3, section 12.1
- Online homework 12.2, questions 1–10

Links to other subjects

- **Science** – to calculate questions involving speed
- **Physical education** – to make calculations involving speed in athletic events

Key words

- No new key words for this topic

Problem solving and reasoning help

- The questions in Exercise 12B of the Pupil Book require pupils to be able to convert between units of measure and then apply their understanding of speed, distance and time to real-life problems.

Common misconceptions and remediation

- A common error made by pupils is to misunderstand the different ways that times (speed and distance) are written. Recap this information for any pupils who may struggle with this.
- Pupils may write 1 hour 30 minutes as 1.3 hours or 1 hour 50 minutes instead of 1.5 hours.

Probing questions

- What makes questions involving speed easy to solve? What makes them difficult to solve?
- Explain why travelling a distance of 60 kilometres in 45 minutes is an average of 80 km/h.
- How can you use the units of speed to help you to solve a problem?
- How can you use information given in a problem to help you work out the units for speed?

Part 1

- Show pupils the following table and ask them to fill it in using different measure of distance, time and speed.

Distance	Time	Speed
Metres	Seconds	Metres per second (mps or m/s)

Part 2

- Again show pupils the formulae for solving problems involving speed, distance and time using the triangle notation. Discuss how pupils can use the formula to calculate the speed, given distance and time:

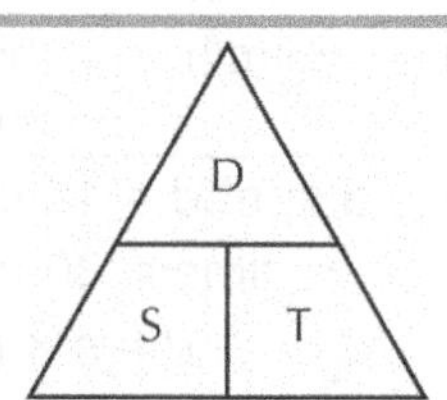

$$\text{speed} = \frac{\text{distance}}{\text{time}}$$

- Go through the following example with pupils:
- The distance between two towns is 495 kilometres. A lorry driver takes 5 hours to travel the first 225 kilometres.
  - What was the lorry driver's average speed for the first part of the journey? (Average speed = distance divided by time, so: 225 km ÷ 5 hours = 45 kmph)

- If the lorry driver's average speed is the same for the whole journey, how long does the journey take? (Time = distance divided by speed, so: 495 miles ÷ 45 kmph = 11 hours)

- **Pupils can now do Exercise 12B from Pupil Book 3.1.**

Part 3

- Go through the formulae linking speed, distance and time. Reinforce the triangle method with pupils.

Answers

Exercise 12B

1 **a** $s = 300 \div 4 = 75$ km/h **b** $s = 300 \div 5 = 60$ km/h **c** $s = 300 \div 6 = 50$ km/h
2 **a** 4 km/h **b** 10 km/h **c** 30 km/h **d** 50 km/h
3 800 km/h
4 40 km/h
5 1080 million km/h
6 6 km/h
7 **a** 40 km/h **b** 20 km/h **c** The speed is halved.

Challenge: Road signs in France

A 19 mph
B 31 mph, 56 mph
C 81 mph, 69 mph

Lesson 12.3 Time

Learning objective

- To work out the time an object will take on a journey, given its speed and the distance travelled

Links to other subjects

- **Science** – to calculate questions involving time
- **Physical education** – to make calculations involving time in athletic events

Resources

- Pupil Book 3.1, pages 199–201
- Intervention Workbook 3, pages 73–76
- Online homework 12.3, questions 1–10

Key words

- No new key words for this topic

Problem solving and reasoning help

- The challenge activity at the end of Exercise 12C in the Pupil Book helps to reinforce the triangle method introduced in the previous two lessons. The triangle method can be used as Part 3 of this lesson.
-

Common misconceptions and remediation

- Pupils will often struggle to convert between fractions, decimals and hours and minutes in time calculations. Use Part 1 to explain and reinforce the correct methods.

Probing questions

- What makes questions involving time easy to solve? What makes them difficult to solve?

Part 1

- Put this table on the board. Then ask pupils to copy and complete it.

Time (hours and minutes)	Time (decimal)	Time (fraction)
1 hour 30 minutes		
	1.75 hours	
		$2\frac{1}{4}$ hours
	1.8 hours	
1 hour 20 minutes		

Part 2

- Again show pupils the formulae for solving problems involving speed distance and time using the triangle notation. Discuss how pupils can use the formula to calculate the time given distance and speed

$$\text{time} = \frac{\text{distance}}{\text{speed}}$$

D

S T

- Go through the following example with pupils:
- A cyclist travels 90 km. Work out the time taken to complete the journey, if the speed of the cycle is: **a** 30 km/h **b** 45 km/h **c** 60 km/h.
- **a** time = distance ÷ speed = 90 ÷ 30 = 3 hours
- **b** time = 90 ÷ 45 = 2 hours
- **c** time = 90 ÷ 50 = 1 hour 30 minutes = 2 hours
- **Pupils can now do Exercise 12C from Pupil Book 3.1.**

Part 3

- Review all the formulae linking speed, distance and time, which pupils have met in this chapter. Reinforce the triangle method with pupils. Ask pupils to write down the formulae straight from the triangle.

Answers

Exercise 12C

1 **a** $t = 40 \div 10 = 4$ hours **b** $t = 40 \div 20 = 2$ hours **c** $t = 40 \div 16 = 2\frac{1}{2}$ hours

2 **a** 5 hours **b** 4 hours **c** 3 hours **d** $3\frac{1}{2}$ hours

3 **a** 3 hours **b** $4\frac{1}{2}$ hours

4 **a** 3 hours **b** $2\frac{1}{2}$ hours

5 12:30 pm

6 **a**

	0	1	2	3
	0	12	24	36

b

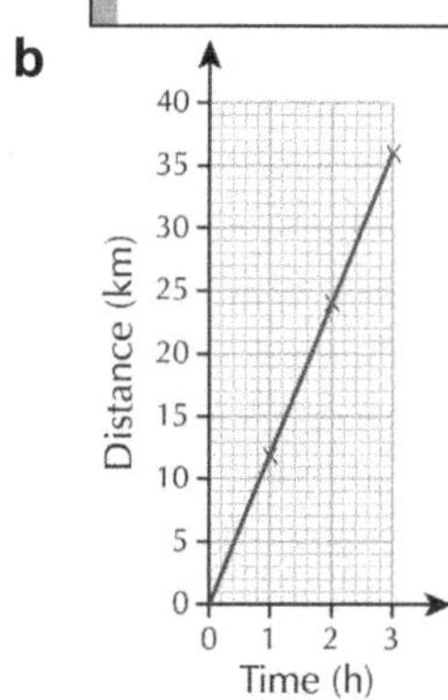

7 10 seconds

Challenge: The distance, speed and time triangle

A 2 hours

B 100 km/h

C 360 km

D $4\frac{1}{2}$ hours

E 100 km/h

F 135 km

Review questions

(Pupil Book pages 202–203)

- The review questions will help to determine pupils' abilities with regard to the material within Chapter 12.
- These questions also draw on the mathematics covered in earlier chapters of the book to encourage pupils to make links between different topics.
- The answers are on the next page of this Teacher Pack.

Financial skills – Shopping at the market

(Pupil Book pages 204–205)

- This financial skills exercise will help pupils to make their learning relevant by applying it to a real-life situation.
- Draw the following table on the board and ask pupils to complete the missing entries.

Kilograms	Grams
0.5 kg	
	2700 g
2.5 kg	
	600 g
$1\frac{3}{4}$ kg	

- Go through the following examples with pupils:
  - 700 g costs £2.10. How much will 100 g cost?
  - 100 g costs $\frac{£2.10}{7}$ = 30p
  - $\frac{3}{4}$ kg of muesli costs £2.40 how much will 2 kg cost
  - $\frac{1}{4}$ kg costs $\frac{£2.40}{3}$ = 80p
  - 2 kg costs 80 × 8 =£6.40
- Which is the better buy: a 24-ounce jar of coffee for £3.69, or a 36-ounce jar of coffee for £4.89?
  - 24-oz jar – 3.69 ÷ 24 = 15p per oz
  - 36-oz jar – 4.89 ÷ 36 = 14p per oz
- The 36-ounce jar is the best buy.
- Pupils can now answer the questions in the Pupil Book.

Answers to Review questions

1 **a** $d = 80 \times 2 = 160$ km **b** $d = 80 \times 3 = 240$ km **c** $d = 80 \times 4\frac{1}{2} = 360$ km

2 **a** $s = 400 \div 5 = 80$ km/h **b** $s = 400 \div 8 = 50$ km/h **c** $s = 400 \div 4 = 100$ km/h

3 **a** $t = 36 \div 6 = 6$ hours **b** $t = 36 \div 3 = 12$ hours **c** $t = 36 \div 4\frac{1}{2} = 8$ hours

4 310 km

5 **a**

Time (h)	0	1	2	3	4	5
Distance (km)	0	5	10	15	20	25

b

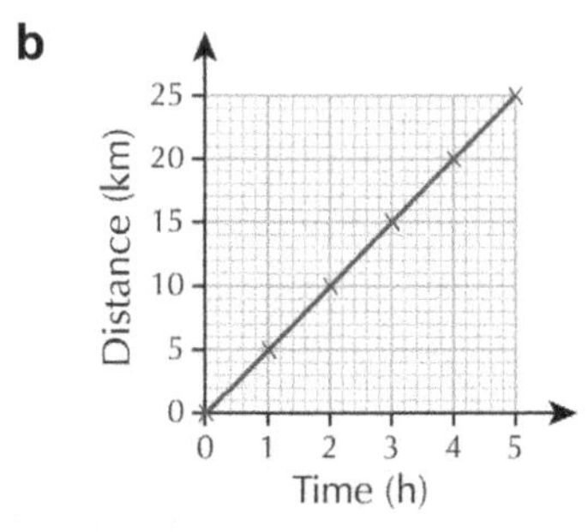

c $12\frac{1}{2}$ km

6 20 km/h

7 1220 km/h

Answers to Financial skills – Shopping at the market

1 Gina

2 **a** For George: 800 g costs £6.00, so 100 g costs $\frac{£6.00}{8}$ = 75 p

b For Natasha: 500 g costs £4.00, so 100 g costs $\frac{£4.00}{5}$ = 80 p

c George pays 75 p for 100 g and Natasha pays 80 p for 100 g, so George gets a better deal.

3 **a** For Mrs Seager: $1\frac{1}{2}$ kg costs £12.00, so $\frac{1}{2}$ kg costs $\frac{£12.00}{3}$ = £4.00

b For Mr Mir: 2 kg costs £14.00, so $\frac{1}{2}$ kg costs $\frac{£14.00}{4}$ = 3.50

c Mrs Seager pays £4.00 for $\frac{1}{2}$ kg and Mr Mir pays £3.50 for $\frac{1}{2}$ kg, so Mr Mir gets a better deal.

4 **a** For Nathan: 1 kg costs 90 p
b For Lily: 1 kg costs 80 p
c So Lily gets a better deal.

5 **a** £1.30 and £1.20 **b** the second one

6 the large tin, as 3×59 p = £1.77

13 Similar triangles

Learning objectives

- What similar triangles are
- Patterns you can find in similar and right-angled triangles
- How to use these patterns to solve some problems

Prior knowledge

- What a right-angled triangle is
- How to use a protractor and construct triangles
- What a hypotenuse is
- How to multiply fractions

Context

- Part of mathematics looks at the relationships between the sides and angles of triangles. In this chapter, pupils will look at important properties of right-angled triangles and learn about similar triangles. They will apply their learning to real-life problems and be in a more secure position to move on to trigonometry in Key Stage 4.

Discussion points

- Is it possible to draw a triangle with:
    - one acute angle
    - two acute angles
    - one obtuse angle
    - two obtuse angles?
- Give an example of the three angles if it is possible. Explain why, if it is not possible.
- Explain why a triangle cannot have two parallel sides.

Associated Collins ICT resources

- Chapter 13 interactive activities on Collins Connect online platform
- *Landing on an aircraft carrier* and *Trig slopes* Wonder of Maths on Collins Connect online platform (note that these may only be suitable for **more able** pupils)

Curriculum references

Reason mathematically

- Extend and formalise their knowledge of ratio and proportion in working with measures and geometry, and in formulating proportional relations algebraically

Geometry and measures

- Use Pythagoras' theorem and trigonometric ratios in similar triangles to solve problems involving right-angled triangles

Fast-track for classes following a 2-year scheme of work

- This chapter is new material and in many cases quite complex. Choose examples carefully to support or challenge pupils.

Lesson 13.1 Similar triangles

Learning objective

- To understand what similar triangles are

Links to other subjects

- **Design and technology** – to calculate unknown heights in 3D layouts
- **Physical education** – to calculate unknown distances and heights for projectiles
- **Geography** – to calculate distances on a map

Resources

- Pupil Book 3.1, pages 207–209

Key words

- denominator
- numerator

Problem solving and reasoning help

- This lesson guides pupils through an extended exploration of the constant relationships between the sides in similar right-angled triangles. Encourage discussion and questioning by pupils. Make sure that you emphasise the key message that the ratios of the sides are constant in similar triangles, as this paves the way for work on trigonometry in Key Stage 4.

Common misconceptions and remediation

- Pupils sometimes learn a set of rules that seem quite complex because they do not make the link to similar triangles and ratios. This chapter gives pupils the opportunity to explore these relationships before formalising the rules of trigonometry at a later stage.

Probing questions

- Which of these statements is/are true? Explain your reasoning.
    - Any two right-angled triangles will be similar.
    - If you enlarge a shape you get two similar shapes.
- What do you look for when deciding if two triangles are similar?

Part 1

- Put this definition on the board:
  Two triangles are similar if they both have the same angles.
- Below this, draw some similar triangles on the board for pupils to match.
- To make it a little more challenging, leave out angles that pupils have to work out using their knowledge of angle facts in a triangle.

Part 2

- Tell pupils that every right-angled triangle obeys the rule of Pythagoras' theorem.
- In about the fourth or fifth century BC, mathematicians discovered something very interesting about the sides and angles in similar right-angled triangles.
- Say that Exercise 13A in this lesson in the Pupil Book is an extended investigation that will lead pupils to find out what the mathematicians discovered.
- **Pupils can now do Exercise 13A from Pupil Book 3.1.**

Part 3

- Ask pupils to summarise what they have learnt *before* they read the summary on page 210 of the Pupil Book.
- Then let pupils read the summary in pairs to identify any outstanding questions. Make sure that the relationship between similar triangles and ratio is secure before moving on to the next lesson. Revisit this at the start of the next lesson.

Answers

Exercise 13A

1 a

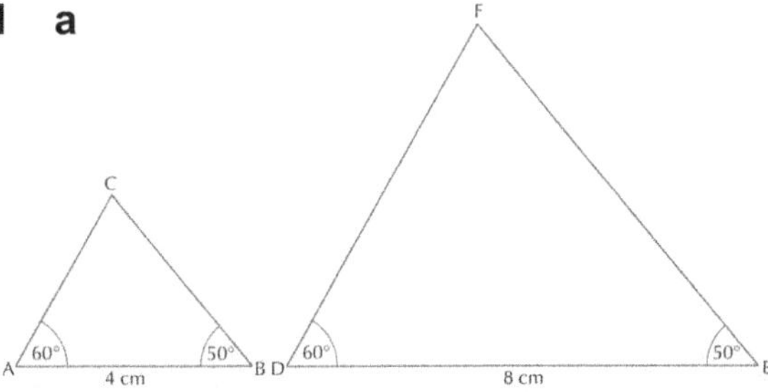

b

c $\frac{DE}{AB} = \frac{8}{4} = 2$, $\frac{EF}{BC} = \frac{7.4}{3.7} = 2$, $\frac{DF}{AC} = \frac{6.6}{3.3} = 2$

d All the answers are equal to or very close to 2.

2 a

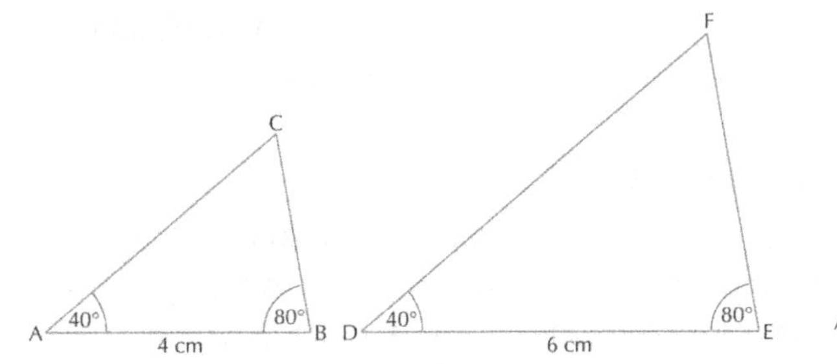

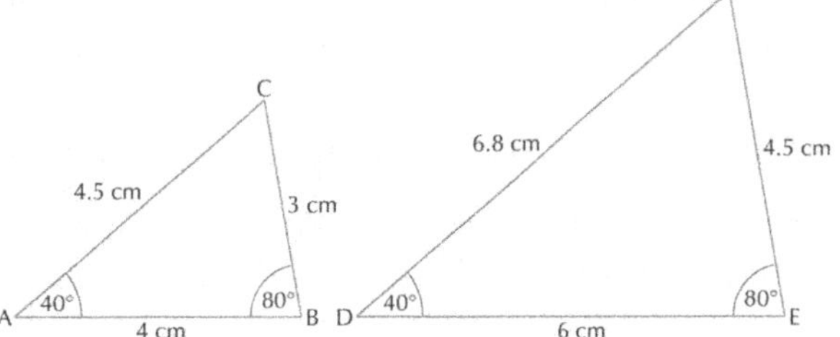

c $\frac{DE}{AB} = \frac{6}{4} = 1.5$, $\frac{EF}{BC} = \frac{4.5}{3} = 1.5$, $\frac{DF}{AC} = \frac{6.8}{4.5} = 1.5$

d All the answers are equal to, or very close to 1.5.

3 a, b

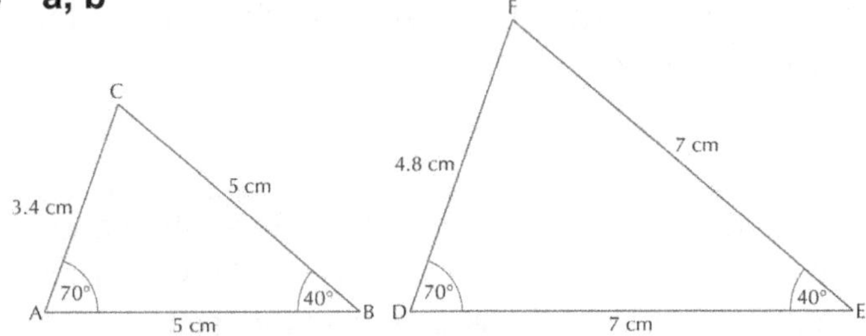

c $\frac{DE}{AB} = \frac{7.5}{1.4} = 1.4$, $\frac{EF}{BC} = \frac{7}{5} = 1.4$, $\frac{DF}{AC} = \frac{4.8}{3.4} = 1.4$

d All the answers are equal to or very close to 1.4.

4 a–c pupils' own answers

d All answers in part c should be the same or very close to each other.

Investigation: Angles and ratios

Pupils' own answers – they should find that they get the same value for all 6 triangles in each of the last 2 columns.

Lesson 13.2 A summary of similar triangles

Learning objective

- To understand what similar triangles are

Resources

- Pupil Book 3.1, pages 210–213
- Online homework 13.2, questions 1–10

Links to other subjects

- **Design and technology** – to calculate unknown heights in 3D layouts
- **Physical education** – to calculate unknown distances and heights for projectiles
- **Geography** – to calculate distances on a map

Key words

- No new key words for this topic

Problem solving and reasoning help

- This lesson consolidates what pupils learnt in Lesson 13.1. **PS** questions 3 to 5 in Exercise 13B of the Pupil Book give pupils the opportunity to apply their learning to some real-life problems. The investigation takes another step towards the later introduction of trigonometry. This is very important, and in order to stress this key concept, you may want to discuss possible solutions to these questions as a class before moving on. The investigation is also an example of geometrical reasoning, which you may want to approach as a class or with guided group discussion, depending on how pupils have responded so far.

Common misconceptions and remediation

- Pupils sometimes learn a set of rules that seem quite complex because they do not make the link to similar triangles and ratios. This lesson formalises the relationships that pupils explored independently in Lesson 13.1. This chapter takes time to prepare pupils working at this level for later work on trigonometry. Try to stress the simplicity of the idea.

Probing questions

- Explain how, to find the missing sides, you can use the fact that these triangles are similar.
- Design your own real-life question using similar triangles.

Part 1

- Pupils should work in pairs to revisit what they have learnt in Lesson 13.1. Then they should work independently to summarise in their own words what they discovered in Lesson 13.1.
- Share some examples, discussing the strengths and any mistakes or weaknesses.
- Give pupils a few minutes to amend their own summaries if they would like to do so.

Part 2

- This lesson builds carefully towards an understanding of similar triangles that can later be used to introduce trigonometry. Therefore, it is important for pupils to consolidate their understanding carefully and then have the opportunity to apply it to a range of problems. Remind pupils that they learned in Lesson 13.1 and Part 1 of his lesson, that for any two similar triangles ABC and DEF, the ratios of the corresponding sides are always the same.
  $\frac{DE}{AB} = \frac{EF}{BC} = \frac{DF}{AC}$
- Tell pupils that they can use these ratios to solve problems in the classroom *and* in real life.

- Work through Example 1 on page 210 of the Pupil Book as a class on the board.
- Use pair share to give pupils time to consider how they might use the ratios above, applied to this problem, before demonstrating.
- **Pupils can now do Exercise 13B from Pupil Book 3.1.**

Part 3

- The investigation at the end of Exercise 13B would make an excellent plenary to this lesson.

Answers

Exercise 13B

1 **a** $\frac{DE}{AB} = \frac{EF}{BC} = \frac{DF}{AC}$ **b** $\frac{GH}{JK} = \frac{HI}{KL} = \frac{GI}{JL}$

c $\frac{QR}{MN} = \frac{RS}{NP} = \frac{QS}{MP}$ **d** $\frac{XY}{TV} = \frac{YZ}{VW} = \frac{XZ}{TW}$

2 **a** FE = 7.5 cm, AC = 2.4 cm **b** KL = 3.2 cm, GH = 6.25 cm

c QS = 6 cm, PN = 6.25 cm **d** YZ = 25 cm, TW = 8.4 cm

3 18 m

4 40 m

5 120 m

Investigation: Nested triangles

A $\angle A = \angle B$, $\angle E = \angle D$ and both triangles share $\angle C$, so the triangles are similar.

B AB = 12.5 − 5 = 7.5 cm

Lesson 13.3 Using triangles to solve problems

Learning objective

- To understand that triangles can be used to solve some real-life problems

Resources

- Pupil Book 3.1, pages 214–215
- Online homework 13.3, questions 1–10

Links to other subjects

- **Design and technology** – to calculate unknown heights in 3D layouts
- **Physical education** – to calculate unknown heights for projectiles
- **Geography** – to calculate unknown bearings on a map

Key words

- No new key words for this topic

Problem solving and reasoning help

- All the questions in Exercise 13C of the Pupil Book are **PS**, and give pupils the opportunity to apply their learning to real-life contexts. Question 4 requires pupils to draw their own diagram. You may want to discuss the use of diagrams before pupils tackle these questions. The investigation at the end of the exercise is a rich activity designed to consolidate learning. Encourage **more able** pupils to prove their findings algebraically and present a justification to the class if appropriate. You could use a dynamic software package to help pupils.

Common misconceptions and remediation

- Pupils may learn a set of rules that seem complex because they do not make the link to similar triangles, ratio and inverse relationships. This chapter takes time to prepare pupils working at this level for later work on trigonometry. Try to stress the simplicity of the idea.

Probing questions

- 'You can use similarity to find missing angles in all triangles.' Is this statement true? How might you adapt it to make it more accurate?

Part 1

- Draw sets of similar triangles on the board but mix them up. Use as many different orientations as possible.
- Mark one angle on one triangle in each set. In each case, ask pupils to identify the similar sets and the marked angle.

Part 2

- Tell pupils that based on the work they have done so far in this chapter they should be starting to realise that we can solve many real-life problems by applying some mathematics. In this lesson they will see some more complex examples of how to draw similar triangles, to model the real situation, and use them to find the solution to a problem.
- Work through Example 3 on page 214 of the Pupil Book as a class, on the board.
- **Pupils can now do Exercise 11C from Pupil Book 3.1.**

Part 3

- Write a set of word problems that involve similar triangles, on the board. Ask pupils to sketch a suitable diagram for each problem. Then ask pupils to identify the information that they need to work out for each question, and set out the mathematical equation they will need to solve.
- Share solutions, discussing any tips that might be useful when decoding word problems. These may be specific, for example, identifying similar triangles, or transferable, for example, text marking to highlight important pieces of information.

Answers

Exercise 11C

1 284 cm
2 328 m
3 82 m
4 A ship sails on a direction of N75°E for 150 km.

a

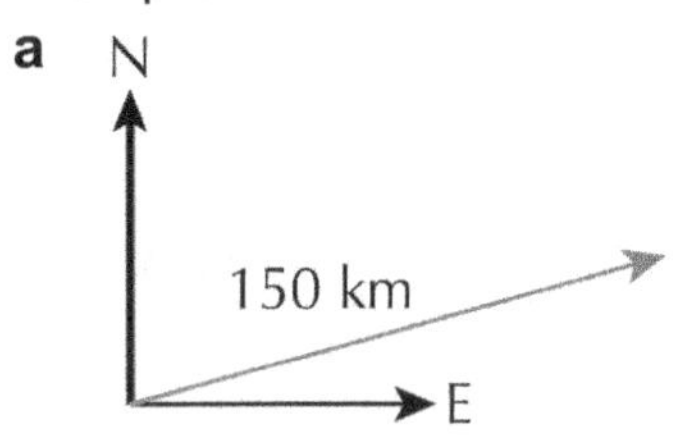

b 145 km

5 4 m
6 **a** 150 km **b** 260 km

Investigation: Skewed triangles

A ∠A = ∠E, ∠B = ∠D and ∠C is the same for both triangles, so they are similar.
B AE = 14.45 cm, BD = 19.43 cm

Review questions

(Pupil Book pages 216–217)

- The review questions will help to determine pupils' abilities with regard to the material within Chapter 13.
- These questions also draw on the mathematics covered in earlier chapters of the book to encourage pupils to make links between different topics.
- The answers are on the next page of this Teacher Pack.

Investigation – Barnes Wallis and the bouncing bomb

(Pupil Book pages 218–219)

- This investigation is an interesting application of the learning in this unit. Pupils may be familiar with the idea from films but will probably be surprised at its use here.
- This is a good opportunity to demonstrate links to other subjects, in this case history.
- As a warm-up to this activity you could show a clip from the original 'Dambusters' movie such as the one at this link: **https://www.youtube.com/watch?v=JM1VGw0wM7k**
- Ask pupils what they know about the story and what bits of information they think the designers and pilots would need, in order to hit their targets.
- Pupils can now work on the investigation questions individually or in groups.
- Encourage **more able** pupils to research other interesting and less-familiar uses of what they have learnt in this lesson.

Answers to Review questions

1 **a** $\frac{DE}{AB} = \frac{EF}{BC} = \frac{DF}{AC}$ **b** $\frac{GH}{JK} = \frac{GI}{JL} = \frac{HI}{KL}$

c $\frac{MN}{QR} = \frac{MP}{QS} = \frac{NP}{RS}$ **d** $\frac{XZ}{TW} = \frac{YZ}{VW} = \frac{XY}{TV}$

2 **a** DE = 8 cm, BC = 7.5 cm **b** JK = 5.6 cm, HI = 6.25 cm

c QR = 7.33 cm, NP = 7.5 cm **d** XY = 22.4 cm, TW = 2.81 cm

3 73.3 m

4 869 m

5 **a** He could sketch an accurate similar triangle and use the measurements from this to work it out.

b 14.6 m

6 **a** 73° **b** 73.6 m

Answers to Investigation – Barnes Wallis and the bouncing bomb

1

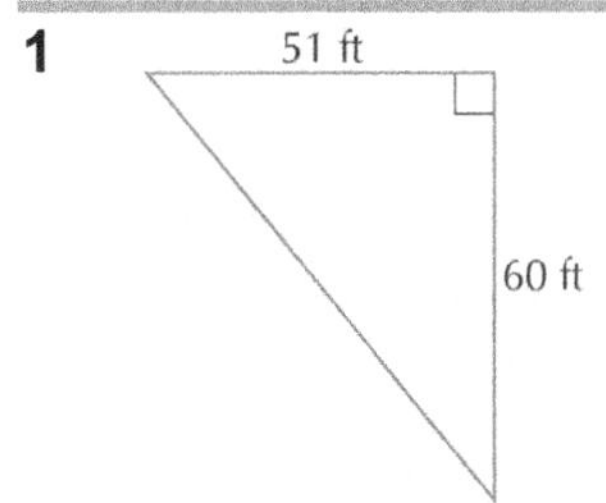

2 50°

3 56°

4 63°

5 **a** Check pupils' drawings. **b** 4 cm **c** 400 m

6 17 cm

14 Revision and GCSE preparation

This chapter is going to

- Help pupils practise and revise topics covered in their current course
- Get pupils started on their GCSE course

Context

- The exercises in this chapter of the Pupil Book cover the following mathematical strands:
  - Alegbra
  - Statistics
  - Geometry and measures
  - Number
- The material will provide excellent practice so that pupils become mathematically fluent. Encourage pupils to work through this whole chapter before their End of Year 9 tests.

Practice

Pupil Book pages 221–235

- The practice exercises focus on key areas, which have been selected from the Pupil Book, so that pupils can consolidate their understanding before moving on to GCSE. Each strand is treated separately, so teachers can select the appropriate areas for pupils to practise.

PB page	Exercise	Exercise provides:
221	14A	Practice in fractions, decimals and percentages
223	14B	Practice in the four rules, ratios and directed numbers
225	14C	Practice in basic rules of algebra and solving linear equations
227	14D	Practice in graphs
229	14E	Practice in geometry and measures
232	14F	Practice in statistics and probability

Revision

Pupil Book pages 235–254

- Exercises 14G to 14Q cover key areas of numeracy that pupils need to grasp, so that they can fully understand algebra and equations. Handling data, including probability, and areas of shape and space are also covered. Pupils often struggle with these important areas of mathematics, so the questions should help to improve their understanding and fluency before they move on to GCSE. Encourage pupils to work through all the questions. Before pupils work on each exercise, go through the examples.

PB page	Exercise	Exercise covers:
235	14G	Revision of BIDMAS
236	14H	Revision of adding and subtracting negative numbers
238	14I	Revision of multiples, factors and prime numbers
240	14J	Revision of squares, square roots and powers
241	14K	Revision of decimals in context: addition and subtraction
243	14L	Revision of decimals in context: multiplication and division
244	14M	Revision of long multiplication and long division in real-life problems
246	14N	Revision of geometry
248	14O	Revision of symmetry
249	14P	Revision of statistics and statistical techniques
252	14Q	Revision of probability

GCSE-type questions

Pupil Book pages 254–257

- The GCSE-type questions in the final section of this chapter are questions that cover some of the topics taught in this book. The questions could be used at the end of Year 9 or at the start of GCSE. Teachers can choose the topics that they feel are appropriate for their pupils.

Note: Answers to all questions in this chapter start on the next page of this Teacher Pack.

Answers to Practice

1 **a** half **b** less than a third **c** more than a quarter
2 **a** 567 **b** 161
3 **a** 5.19 m **b** 22.65 km
4 $\frac{7}{12}$
5 £55.50
6 134.4
7 **a i** about 25% **ii** about 45%
b i about 30% **ii** about 60%
8 82 kg
9 £106.60.
10 60%, 0.6, $\frac{3}{5}$, $\frac{6}{10}$
11 **a** $\frac{14}{15}$ **b** $\frac{7}{18}$ **c** $4\frac{3}{20}$
12 £34

Exercise 14B

1 **a** 471 **b** 379 **c** 264 **d** 22
2 **a i** 50 **ii** 48
b +12
3 **a** 440 **b** 860 **c** 8 **d** 401 **e** 297
f 311
4 **a i** £3.95 **ii** £6.05
b no, 15 p short
5 **a** 9 + 6 = 20 – 5
b 15 – 3 = 4 × 3
c 5 × 2 = 15 – 5 or 5 – 2 = 15 ÷ 3
d 8 ÷ 4 = 4 – 2 or 8 ÷ 4 = 4 ÷ 2
6 £31.36
7 **a** −6 and 5 or −4 + 3 or −2 + 1 **b** −11 **c** −8 – 5 = −13
8 **a** 508.4 **b** 2.5 **c** 12.5 **d** 10.4
9 17 bins (with £7 left over)
10 £18
11 **a** 15 **b** £56 175 **c** £80.25
12 **a** 52 mph **b** 4 hours 30 minutes

Exercise 14C

1 **a** £70 **b** 4 hours
2 **a** $x = 2$ **b** $x = 4$ **c** $x = 16$
3 **a** $6x$ **b** £$5y$ **c** £xy
4 **a** (10, 9) **b** The first number has to be even.
5 **a** $x + 4$ **b** $y - 2$
6 **a** $b = 2a$, $\frac{a+b+c}{3} = 35$, $a = 30$, $b + c = 75$ **b** 60 kg and 15 kg
7 **a**

	n	4
n	n^2	$4n$
4	$4n$	16

b $n^2 + 8n + 16$
8 **a** $4x - 20$ **b** $11x + 3$ **c** $5x + 2$ **d** $17x + 16$ **e** $5x + 22$

9 **a i** 21 **ii** 10 **iii** 50
b i $z = 3$ **ii** $z = 22$ **iii** $z = -1$
10 $6x + 3 = 12$, $x = 1.5$

Exercise 14D

1 **a** cross added at (4, 5) **b** (3, 5), (4, 5), (5, 5) and (6, 5)
c The *y*-coordinate is always 5.
2 **a** (3, 3) **b** (1, 1)
3 **a** A(2, 1) and B(0, 3) **b** (5, 4)
4

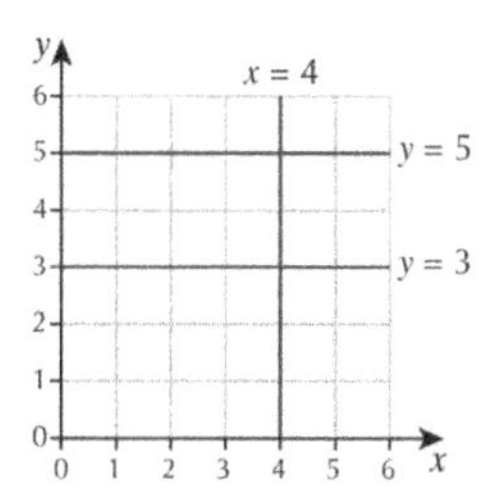

5 **a** l_3 **b** l_4 **c** l_2 **d** l_1
6 **a** 10 minutes **b** 3.5 miles
7 **a**

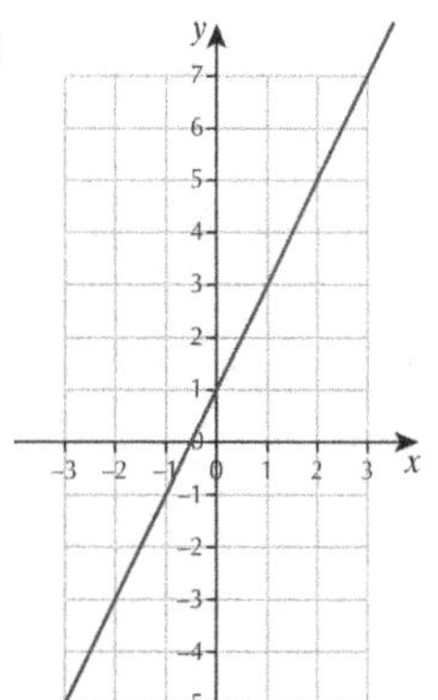

b

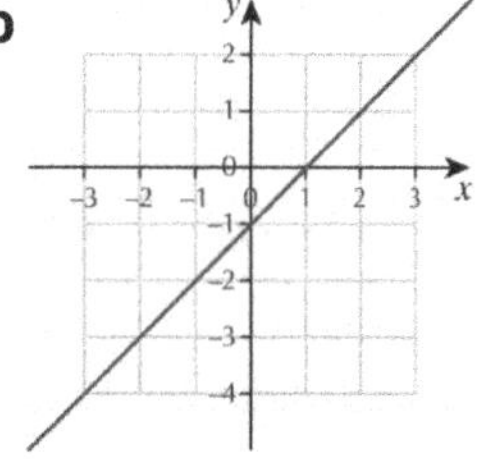

c

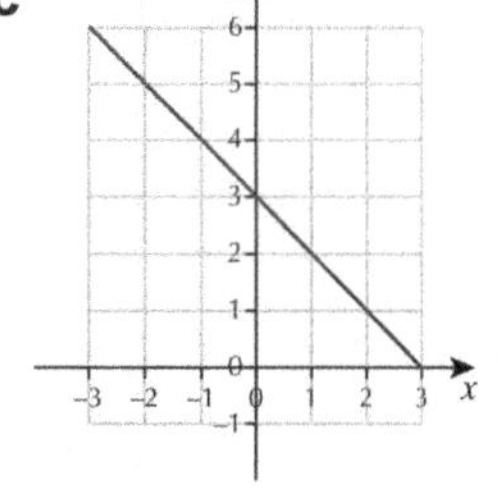

8 yes, 2 × 20 – 10 = 30
9 **a ii** and **iv**
b

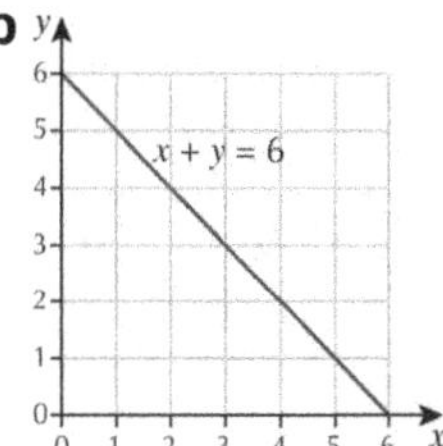

Exercise 14E

1 **a i** 2 **ii** 2
b i 6 **ii** 6
c i 0 **ii** 2
2 **a i** rectangle **ii** kite **iii** parallelogram
b i 4 **ii** 2 **iii** 2
3 **a** 21 cm^2 **b** 7.5 cm^2 **c** same area as the rectangle, 21

4 **a** *a* – acute, *b* – obtuse, *c* – acute, *d* – right-angled, *e* – reflex and *f* – obtuse
b smaller because CD is parallel to AB and CB is not parallel to AF

5 **a**

		Number of lines of symmetry				
		0	1	2	3	4
Order of rotational symmetry	1		A			
	2	F		B, C, D		
	3				G	
	4					E

b rectangle or rhombus

6 **a** Check pupils' diagrams. **b** 82°

7 a = 54°, b = 82°, c = 152°

8 **a** 80 **b** No, it is just over 60 mph.
c 50 km is approximately 31 miles, so 150 km is approximately 93 miles.

9 **a** 288 cm^2 **b** 4 **c** 16 : 1

Exercise 14F

1 **a** 13 **b** 4 **c** 18

2 **a** 7 **b** 11 **c** 2

3 **a** $\frac{1}{2}$
b (H, 1), (H, 2), (H, 3), (H, 4), (H, 5), (H, 6), (T, 1), (T, 2), (T, 3), (T, 4), (T, 5), (T, 6)
c $\frac{1}{6}$

4 **a** 2 **b** 3 **c** 4 **d** 6

5 **a** Q **b** R **c** P and R, angles are the same

6 **a**

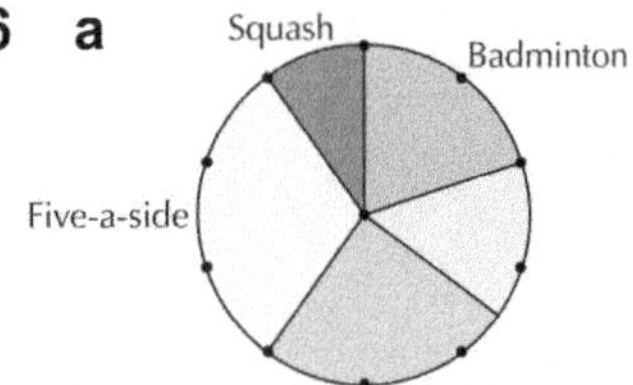

b **i** 25% **ii** 15%
c **i** 54 **ii** 36
d 30% of 180 = 54 and 20% of 280 = 56

7 **a** $\frac{3}{16}$ **b** $\frac{3}{8}$
c **i**

		Score on first die			
		1	2	3	4
Score on second die	1	1	2	3	4
	2	2	4	6	8
	3	3	6	9	12
	4	4	8	12	16

ii 1

8 two numbers in the ratio 2 : 1, for example: 20 red and 10 blue

9 0.3

Exercise 14G

1 **a** 25 **b** 10 **c** 12 **d** 12 **e** 12
f 37

2 **a** 30 **b** 3 **c** 9 **d** 12 **e** 4
f 12

3 **a** 8 **b** 49 **c** 11 **d** 3 **e** 3
f 32 **g** 13 **h** 23 **i** 4 **j** 50
k 48 **l** 20

4 **a** $4 \times (3 + 7) = 40$ **b** $10 \div (2 + 3) = 2$ **c** $18 \div (3 + 3) = 3$
d $(5 - 2) \times 4 = 12$ **e** $(20 - 5) \times 2 = 30$ **f** $5 \times (12 - 8) = 20$
g $(10 - 2^2) \times 2 = 12$ **h** $10 - (2^2 \times 2) = 2$ **i** $(20 - 4^2) \times 5 = 20$

5 **a** 30 **b** 22 **c** 21 **d** 12 **e** 7
f 81

Exercise 14H

1

	Amount paid in		
			£64.37
Standing order		£53.20	£11.17
Cheque	£32.00		£43.17
Direct debit		£65.50	–£22.33
Cash	£20.00		–£2.33
Wages	£124.80		£122.47
Loan		£169.38	–£46.91

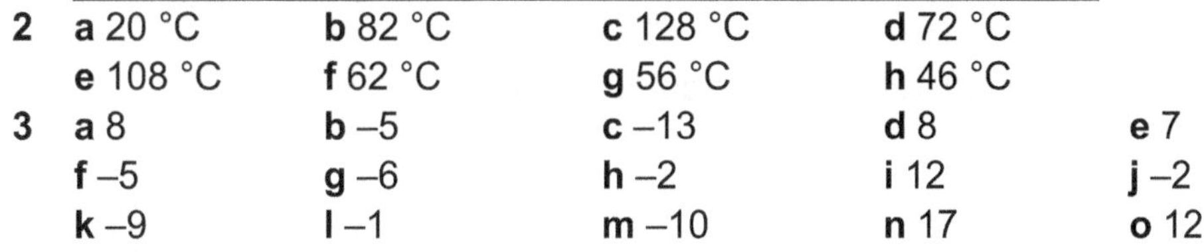

2 **a** 20 °C **b** 82 °C **c** 128 °C **d** 72 °C
e 108 °C **f** 62 °C **g** 56 °C **h** 46 °C

3 **a** 8 **b** –5 **c** –13 **d** 8 **e** 7
f –5 **g** –6 **h** –2 **i** 12 **j** –2
k –9 **l** –1 **m** –10 **n** 17 **o** 12
p –16

4

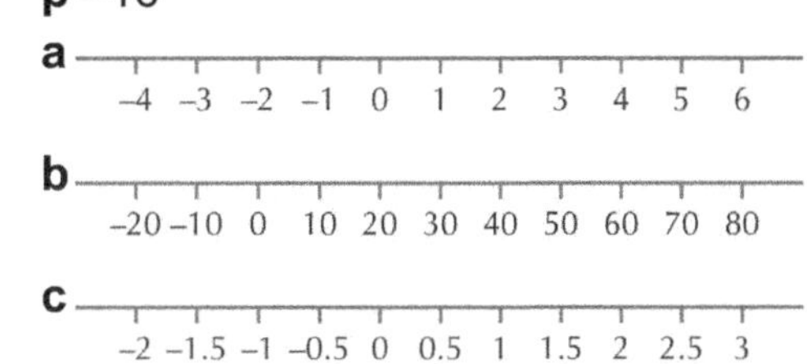

5 **a** –2 **b** 4 **c** –3 **d** –2 **e** –9
f 3 **g** –15 **h** 4 **i** 7

Exercise 14I

1 **a** 4, 8, 12, 16, 20 **b** 9, 18, 27, 36, 45 **c** 12, 24, 36, 48, 60
d 25, 50, 75, 100, 125

2 **a** 3, 15, 18, 24, 36, 39, 45, 48, 69, 90, 120 **b** 15, 45, 90, 120
c 8, 24, 36, 48, 64, 120 **d** 24, 36, 48, 120

3 **a** 48 **b** 48 **c** 45 **d** 45

4 **a** 2, 3, 5, 7, 11, 13, 17, 19 **b** prime numbers

5 **a** 1, 2, 3, 4, 6, 8, 12, 16, 24, 48 **b** 1, 2, 4, 13, 26, 52

c 1, 2, 3, 4, 5, 6, 10, 12, 15, 20, 30, 60 **d** 1, 3, 5, 15, 25, 75
e 1, 2, 4, 5, 10, 20, 25, 50, 100 **f** 1, 2, 5, 10, 13, 26, 65, 130

6 The numbers left are: 2, 3, 5, 7, 11, 13, 17, 19, 23, 29, 31, 37, 41, 43, 47, 53, 59. They are prime numbers.

Exercise 14J

1 **a** 49 **b** 81 **c** 121 **d** 169 **e** 225
f 361 **g** 576 **h** 625 **i** 1024 **j** 2809

2 **a** 6 **b** 8 **c** 10 **d** 12 **e** 14
f 6.32 **g** 8.94 **h** 10.95 **i** 22.36 **j** 28.28

3 **a** 1024 **b** 1728 **c** 28 561 **d** 9261

4 **a** 141.421 36 **b** 447.2136 **c** 1414.2136

5 **a i** 1 **ii** 1 **iii** –1 **iv** –1
b −1, as the power is an odd number

Exercise 14K

1 £4359.33
2 £11.22
3 12.9 cm
4 0.590 kilograms
5 £252.08
6 £9000.61
7 7.5 cm
8 £484.83

Exercise 14L

1 **a** £106.80 **b** £72.84 **c** £1908 **d** £1023.30
2 £37.35
3 £20.23
4 730.8 cm
5 £2312.20
6 £125.80
7 £69.75
8 £246.60

Exercise 14M

1 810 words
2 36
3 **a** 43 buses **b** 33 000
4 **a** 814 **b** £5.60
5 **a** 58 **b** 4
6 990 grams
7 **a** £255 **b** 250 **c** £50.21

Exercise 14N

1 **a i** 12 cm **ii** 9 cm^2
b i 18 cm **ii** 20 cm^2
c i 44 mm **ii** 120 mm^2
d i 34 m **ii** 60 m^2

2 **a** 4 cm^2 **b** 20 cm^2 **c** 300 mm^2 **d** 14 m^2

3 **a** 66 cm^2 **b** 96 cm^2 **c** 12 m^2 **d** 80 m^2
4 **a i** 62 cm^2 **ii** 30 cm^3
b i 150 cm^2 **ii** 125 cm^3
c i 28 cm^2 **ii** 8 cm^3
5 20 cm^2

Exercise 14O

1 **a** 2 **b** 2 **c** 6 **d** 4 **e** 5
2 **a** 2 **b** 1 **c** 1 **d** 4
3 **a** 2 **b** 2 **c** 5 **d** 4 **e** 2
4 **a** 4 **b** 3 **c** 4 **d** 2
5 **a** 1
b
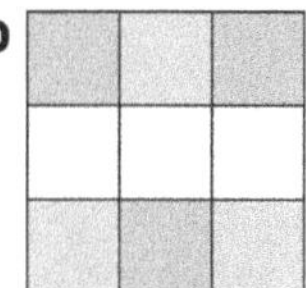

Exercise 14P

1 **a i** 1 **ii** 4 **iii** 8
b i 5 **ii** 6 **iii** 6
c i £4.50 **ii** £3.25 **iii** £3.40
d i 18 **ii** 20 **iii** 21
2 **a**

	Boys		Girls	
	Tally	Frequency	Tally	Frequency
12 ≤ T < 14	||	2	||	2
14 ≤ T < 16	||||	4	|	1
16 ≤ T < 18	|||	3	||	2
18 ≤ T < 20		0	||||	4
20 ≤ T < 22	|	1	|	1

b 14 ≤ T < 16
c 18 ≤ T < 20
3 as you get older, the longer it takes to finish, or good positive correlation
4 **a**
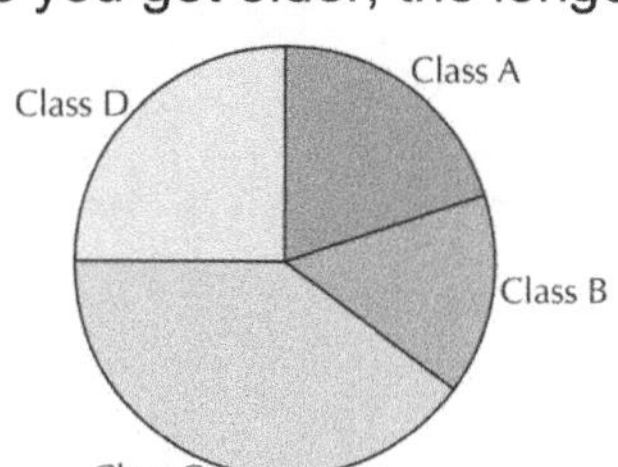

b There may be a different number of pupils in the two classes.

Exercise 14Q

1 **a** $\frac{1}{10}$ **b** $\frac{1}{2}$ **c** $\frac{1}{5}$ **d** $\frac{2}{5}$
2 **a** 4: HH, HT, TH, TT **b** $\frac{1}{4}$ **c** $\frac{1}{4}$ **d** $\frac{1}{2}$
3 **a** $\frac{3}{25}$ **b** $\frac{1}{5}$
c Not fair: you would expect the spinner to land on each number about 10 times.

Answers to GCSE-type questions

1 **a** 27 and 73 **b** 15 and 55 **c** 27 and 72 **d** 42

2 **a** £3.07 **b** 5

3 3 × 1.6 = 4.8 km, so, yes, as 4.7 km < 3 miles.

4 19 656

5 **a** 17 and 81

b **i** 69 **ii** 81

6 **a** multiple **b** square root **c** cube **d** factor **e** square

7 **a** 1, 2, 3, 6, 9, 18 **b** any multiple of 10 **c** 15

d $\frac{2}{5}$

8 **a** 16 °C **b** 4 °C

9 **a** 14, 77, 104, 140, 147 **b** –8, –4, –2, 0, 2

c 0.09, 0.091, 0.9, 0.901, 0.91 **d** $\frac{1}{2}$, 0.55, 60%, $\frac{5}{8}$

10 no, $6^2 = 6 \times 6 = 36$

11 **a**

	Theme Park	Blackpool	Total
Boys	25	18	43
Girls	17	30	47
Total	42	48	90

b **i** $\frac{47}{90}$ **ii** $\frac{8}{15}$

12 175° + 115° + 95° = 385°; the total should be 360°.

13 **a** 33 **b** an odd number

14 **a** £6.00 **b** They work out to be the same total price.

15 **a** 9 feet **b** 4.8 m

c 5 m is just less than 16 feet, so Helen has the higher record.

16 19

17 Check pupils' constructions.

18 **a** 5 **b** 300 cm^2

19 **a** **i** 135 **ii** 60

b 160°

c 105

1 Percentages

Learning checklist

- I can calculate percentages. ☐
- I can change a value by a given percentage. ☐
- I can change fractions to decimals, and decimals to percentages. ☐

- I can find an original amount after a percentage change. ☐

2 Equations and formulae

Learning checklist

- I can expand brackets. ☐
- I can factorise simple algebraic expressions. ☐
- I can substitute into simple formulae. ☐
- I can solve equations that have brackets or fractions or both. ☐

3 Polygons

Learning checklist

- I know the names of different polygons. ☐
- I can recognise regular, irregular, convex and concave polygons. ☐

- I can work out the sum of the interior angles for different polygons. ☐
- I can work out the size of each interior angle for different regular polygons. ☐

4 Using data

Learning checklist

I know how to interpret simple graphs and charts, and how to draw conclusions. ☐
I know how to interpret simple two-way tables. ☐
I know how to compare data from two simple sets of data. ☐

I know how to interpret graphs and charts, and how to draw conclusions. ☐
I know how to interpret a variety of two-way tables. ☐

I know how to draw conclusions from scatter graphs and I have a basic understanding of correlation. ☐

5 Circles

Learning checklist

I can calculate the circumference of a circle. ☐
I can calculate the area of a circle. ☐

6 Enlargements

Learning checklist

I can enlarge a shape by a given scale factor. ☐

I can enlarge a shape about a centre of enlargement. ☐
I can draw rays to find a centre of enlargement. ☐

7 Fractions

Learning checklist

- I can add or subtract two simple fractions. ☐

- I can add or subtract two fractions with different denominators. ☐
- I can multiply two simple fractions. ☐
- I can divide two simple fractions. ☐

8 Algebra

Learning checklist

- I can expand simple brackets. ☐
- I can expand simple brackets and simplify. ☐

- I can factorise simple algebraic expressions. ☐
- I can expand brackets that give rise to powers. ☐

9 Decimal numbers

Learning checklist

- I can round numbers to the nearest 10, 100 or 1000. ☐
- I can multiply and divide by 10, 100 and 1000. ☐
- I can solve simple number problems. ☐

- I can multiply and divide by simple powers of 10. ☐
- I can round numbers correct to one decimal place. ☐

- I can round numbers in order to make sensible estimations. ☐
- I can multiply and divide by any positive power of 10. ☐

10 Surface area and volume of 3D shapes

Learning checklist

- I can work out the surface areas of cubes and cuboids. ☐
- I can work out the volumes of cubes and cuboids. ☐
- I can work out the capacities of cubes and cuboids, measured in litres. ☐
- I can work out the volume of a triangular prism. ☐

11 Solving equations graphically

Learning checklist

I know that a graph from an equation in the form $y = mx + c$ is a straight line. ☐

I can solve problems from data that gives a straight-line graph. ☐
I can solve simple quadratic equations by drawing a graph. ☐
I can solve problems from data that gives a quadratic graph. ☐

12 Distance, speed and time

Learning checklist

- I can use the formula $d = st$ to work out a distance. ☐
- I can use the formula $s = \frac{d}{t}$ to work out a speed. ☐
- I can use the formula $t = \frac{d}{s}$ to work out a time. ☐
- I can read and interpret distance–time graphs. ☐

13 Similar triangles

Learning checklist

	I understand what similar triangles are.	☐
	I can use similar triangles to solve simple problems.	☐

3-year scheme of work

The following scheme of work provides a suggestion for how Pupil Book 3.1 can be taught over the course of one year, as part of a 3-year Key Stage 3 course.

Please note that you can recombine the test questions provided on *Collins Connect* to create new tests if your frequency of assessment differs from that below, or if you wish to combine content from different chapters in your own half-term tests.

This scheme of work is provided in editable Word and Excel format on the CD-ROM accompanying this Teacher Pack.

Chapter	Lesson	No. of hours	Learning objective	Comments/ suggestions
Half-term / Term 1				
1 Percentages	1.1 Simple interest	1	• To understand what simple interest is • To solve problems involving simple interest	Percentage increase and decrease is probably one of the most common uses of mathematics in real life. Everyone meets it in some form or other, even if only in terms of financial capability. Pupils often struggle when they start using percentages that are greater than 100. Using a real-life example will help pupils to overcome this. Start with percentages that pupils can work with comfortably, for example, an explanation based on a shop selling a pair of jeans that cost £30 to make. A 50% profit would mean that the shop sold the jeans for £30 plus £15, which is £45; 100% profit would mean selling the jeans for £30 plus £30, which is £60; 150% profit would mean selling the jeans for £30 plus £45, which is £75. You could also use the following link to find real-life applications of percentage: **http://www.pfeg.org/**
	1.2 Percentage increases and decreases	1	• To calculate the result of a percentage increase or decrease • To choose the most appropriate method to calculate a percentage change	These lessons reinforce the concept of using percentage as an operator. This is an important step to ensure confidence and fluency in pupils, so make sure that you spend time on this lesson. It often helps to make links to fractions as operators.
	1.3 Calculating the original value	2	• Given the result of a percentage change, to calculate the original value	
	1.4 Using percentages	1	• To revise the links within fractions, decimals and percentages • To choose the correct calculation to work out a percentage	This lesson develops and consolidates pupils' understanding of percentages by enabling them to make choices and decisions about the methods they use in a range of contexts.
	Challenge – The Royal Albert Hall	2		This challenge gives pupils the opportunity to extend their learning to a real-life context. All the information pupils will need is provided in the Pupil Book but it is quite complex. Pupils working at this level may find it difficult to access the information they need. This is representative of how they are likely to find information

				presented in real life. Pupils will need to read the questions very carefully to decide what information they need and what mathematical skills to use in each case.
2 Equations and formulae	2.1 Multiplying out brackets	1	• To multiply out brackets	This chapter builds on previously learned algebraic techniques and moves on to more advanced methods of algebraic manipulation. These include: simplifying expressions and expanding brackets, factorising algebraic expressions, solving linear equations involving fractions and using formulae.
	2.2 Factorising algebraic expressions	1	• To factorise expressions	
	2.3 Equations with brackets	1	• To solve equations with one or more sets of brackets	
	2.4 Equations with fractions	1	• To solve equations involving fractions	
	2.5 Formulae	1	• To practise using formulae	
	Financial skills – Wedding day	1		This financial skills activity gives pupils the opportunity to apply the skills they have learned in the chapter to a practical situation that many pupils may experience in the future. The cost formula used is often encountered in GCSE exams, so it is important for pupils have a good grasp of this.
Chapters 1–2 assessment on Collins Connect				
3 Polygons	3.1 Polygons	1	• To know the names of polygons • To know the difference between an irregular polygon and a regular polygon	This chapter builds on pupils' ability to categorise using polygons, which is a good transferable skill across mathematics and beyond. This chapter then introduces pupils to finding the sums of the interior and exterior angles of polygons.
	3.2 Angles in polygons	1	• To work out the sum of the interior angles of a polygon	
	3.3 Interior angles of regular polygons	1	• To work out the sizes of the interior angles in regular polygons	
	Activity – Regular polygons and tessellations	2		This activity is designed to give pupils the opportunity to apply what they have learnt about the characteristics of polygons to tessellations. Pupils will need to apply what they know about angles in polygons. Tessellations were not part of this chapter but pupils should have met the concept before.
Half-term				
Half-term / Term 2				
4 Using data	4.1 Scatter graphs and correlation	1	• To infer a correlation from two related scatter graphs	This chapter picks up the ideas from the material that pupils learned in statistics in previous years. It develops ways to illustrate distributions and how we can use data to explore possibilities as well as to compare them. The chapter culminates in pupils conducting their own
	4.2 Interpreting graphs and diagrams	1	• To use and interpret a variety of graphs and diagrams	

	4.3 Two-way tables	1	• To interpret a variety of two-way tables	investigations, using the ideas from the first part of the chapter. The following link to a video demonstrates an example of the power of statistics: **http://www.gapminder.org/videos/the-river-of-myths/**
	4.4 Comparing two or more sets of data	1	• To compare two sets of data from statistical tables and diagrams	In order to make comparisons between graphs, pupils need to be able to understand what the graph represents, what the axes mean and how to read data from the graph. Make sure pupils are able to do this before they attempt comparisons. You may need to model some examples.
	4.5 Statistical investigations	1	• To plan a statistical investigation	It is important that pupils' data collection sheets are fit for purpose and have been designed to capture all the factors that have a bearing on the investigation.
	Challenge – Rainforest deforestation	1		Talk to pupils about deforestation and the fact that for years, the big rainforests of the world have been reduced and chopped down, while the country or countries concerned benefit from the cleared land and the revenue from the wood obtained from the trees. Note that this challenge has no intention of making any judgement values of the country or countries concerned. Instead, it has been devised to allow pupils to find what the statistics may suggest; in other words, that economic growth can affect the amount of deforestation.
Chapters 3–4 assessment on Collins Connect				
5 Circles	5.1 The formula for the circumference of a circle	1	• To calculate the circumference of a circle	Tell pupils that the circle is probably the most important shape in the universe. It is also the most mysterious. We use a fascinating number that pupils may have heard of, called pi, written as π, which is used to calculate the circumference (perimeter) of a circle. But π cannot be written exactly as a number and its decimal places never end. Encourage pupils to prepare for this chapter by doing their own research on π. Encourage pupils to present their findings to the class.
	5.2 The formula for the area of a circle	2	• To calculate the area of a circle	
	5.3 Mixed problems	2	• To solve problems involving the circumference and area of a circle	
	Financial skills – Athletics stadium	2		This activity is designed to give pupils the opportunity to apply their knowledge to a multi-step real-life problem. The context is familiar, but the activity is presented in a slightly more complex way than pupils may be used to. All the information required to answer the questions is in the text, but pupils will need to read and then think carefully about how they access the information. Remind them to highlight the key information they will need. Tell pupils that they will need to combine their skills not only from this chapter, but also from different areas of mathematics, for example, number.

6 Enlargements	6.1 Scale factors and enlargements	2	• To use a scale factor to show an enlargement	This chapter starts by showing pupils how to enlarge a 2D shape by a positive whole number scale factor. Pupils are then shown how to enlarge a shape using a centre of enlargement before being taught how to use a coordinate grid to enlarge a shape. Using photographs in the lesson may help their understanding.
	6.2 The centre of enlargement	1	• To enlarge a shape about a centre of enlargement	
	6.3 Enlargements on grids	2	• To enlarge a shape on a coordinate grid	
	Problem solving – Photographs	2		This problem-solving activity consolidates topics previously covered on extracting data, area and ratio.
Half-term / Term 3				
7 Fractions	7.1 Adding and subtracting fractions	1	• To add or subtract any two fractions	The number system was originally used simply for counting, and only positive whole numbers were used. Later, the number system was extended to include zero. Zero is very important as a place holder, and in negative numbers and fractions. By now, pupils should have an understanding of the ordinal value of fractions as well as their use as operators. This chapter builds on the Year 8 work on fractions, leading pupils to using fractions to solve real-life problems.
	7.2 Multiplying fractions	1	• To multiply two fractions	
	7.3 Dividing fractions	1	• To divide one fraction by another	
	Problem solving – The 2016 Olympic Games in Rio	2		Pupils apply their understanding of fractions to a topical but more complex problem. Pupils need to work methodically to identify all the information required to answer the questions. Remind them to highlight the key information they will need. Pupils will also need to combine their understanding across fractions, decimals and percentages as well as their understanding of averages.
Chapters 5–7 assessment on Collins Connect				
8 Algebra	8.1 Expanding brackets	1	• To multiply out brackets with a variable outside them	This chapter recalls previous work on algebra and revisits expansion of brackets and collecting like terms. Pupils are also shown how to expand a bracket, and then factorise a bracket that involves powers. Finally, pupils learn how to expand expressions with two brackets.
	8.2 Factorising algebraic expressions	2	• To factorise expressions	
	8.3 Expand and simplify	2	• To expand expressions with two brackets and simplify them	
	Challenge – California gold	1		This challenge activity requires pupils to apply their learning in an unfamiliar context. Introduce it with some recent examples of treasure trove finds from the internet and get pupils to research the current price of gold per gram.
9 Decimal numbers	9.1 Multiplication of decimals	1	• To practise multiplying decimal numbers	The ability to understand place value is the key to being able to use numbers effectively when doing calculations in real life. Share with pupils some of the different aids to calculation that have been developed since we first used numbers to count.
	9.2 Powers of 10	1	• To understand and work with both positive and negative powers of ten	

	9.3 Rounding suitably	1	• To round numbers, where necessary, to a suitable degree of accuracy	For example, skilled abacus users can calculate very quickly and accurately. You could show a video such as the one at the following link (about Japanese children learning super-fast mathematics with the abacus) to demonstrate different approaches to calculations across the world: **http://www.youtube.com/watch?v=6m6s-uIE6LY** Nowadays, computers and calculators can help you to work out the more complicated calculations. Modern calculators can do so much more than simple arithmetic but pupils need to know how to integrate this with a good basic understanding of place value and mental arithmetic. In this chapter, pupils will learn more about the decimal system of counting and they will practise skills in using a calculator.
	9.4 Dividing decimals	1	• To confirm ability to divide with decimals	
	9.5 Solving problems	1	• To solve real-life problems involving multiplication or division	Pupils often struggle to decode word problems to identify the mathematics they need to use. Provide plenty of opportunity for pupils to discuss word problems to identify the mathematics required independently. Encourage pupils to develop a set of transferable strategies such as marking the text to identify key words and information. Discuss how pupils can apply these strategies to a range of problems across mathematics, particularly in real life.
	Mathematical reasoning – Paper	2		This activity uses the context of paper, with which pupils may be very familiar. All the information pupils need is provided in the text in the Pupil Book, but it is quite complex. Pupils will need to read the questions very carefully to decide on the information that they will need and what mathematical skills to use in each case. The questions move freely between fractions and decimals. This is something that pupils need to be comfortable with, which also develops their conceptual understanding of fractions and decimals being ways of expressing parts of a whole.
Half-term				
Half-term / Term 4				
10 Surface area and volume of 3D shapes	10.1 Surface areas of cubes and cuboids	2	• To work out the surface areas of cubes and cuboids	Remind pupils that perimeter, area and volume are used widely in many jobs and professions, from farming to astronomy. Encourage pupils to ask family and friends if they use these units of measure in their work. Pupils could also explore specific jobs on the internet. A good example is the building industry, which is totally dependent on workers being able to measure lengths and calculate areas.
	10.2 Volume formulae for cubes and cuboids	2	• To use a simple formulae to work out the volume of a cube or cuboid • To work out the capacity of a cube or cuboid	
	10.3 Volumes of triangular prisms	2	• To work out the volume of a triangular prism	

	Investigation – A cube investigation	2		Pupils apply their understanding of area to a more complex problem. Pupils need to work methodically and be able to explain their solutions. This is a good transferable skills objective to share with pupils when doing this investigation. Ask pupils to share not only their solutions but also *how* they approached working on the problem.
Chapters 8–10 assessment on Collins Connect				
11 Solving equations graphically	11.1 Graphs from equations in the form $y = mx + c$	1	• To draw a linear graph from any linear equation • To solve a linear equation from a graph	This chapter provides examples of the fact that many equations can arise from real-life situations, and it builds on straight-line graphs and simple quadratics. Pupils are introduced to the idea that while many equations that are used to model real life are difficult to solve by algebraic methods, but they are solved more easily by drawing a graph. You could show video clips such as the one at this link, of a horse jumping: **http://www.youtube.com/watch?v= mgVi78diBwQ**
	11.2 Problems involving straight-line graphs	1	• To draw graphs to solve some problems	
	11.3 Solving simple quadratic equations by drawing graphs	2	• To solve a simple quadratic equation by drawing a graph	
	11.4 Problems involving quadratic graphs	2	• To solve problems that use quadratic graphs	
	Problem solving – Squirrels	2		Pupils often ask why they have to do mathematics that is not familiar to them. Say that using graphs to monitor wildlife is a good example of how mathematics can be used to benefit society.
Holidays				
Half-term / Term 5				
12 Distance, speed and time	12.1 Distance	2	• To work out the distance travelled in a certain time at a given speed • To use and interpret distance–time graphs	This chapter teaches pupils how to calculate with different measures. Pupils are introduced to the relationship between speed, distance and time. Pupils learn how to solve problems involving these measures.
	12.2 Speed	2	• To work out the speed of an object, given the distance travelled and the time taken	
	12.3 Time	2	• To work out the time an object will take on a journey, given its speed and the distance travelled	
	Financial skills – Shopping at the market	1		This financial skills exercise will help pupils to make their learning relevant by applying it to a real-life situation.
13 Similar triangles	13.1 Similar triangles	2	• To understand what similar triangles are	Part of mathematics looks at the relationships between the sides and angles of triangles. In this chapter, pupils will look at important properties of right-angled triangles and learn about similar triangles. They will apply their learning to real-life problems and
	13.2 A summary of similar triangles	1	• To understand what similar triangles are	

	13.3 Using triangles to solve problems	2	• To understand that triangles can be used to solve some real-life problems	be in a more secure position to move on to trigonometry in Key Stage 4.
	Investigation – Barnes Wallis and the bouncing bomb	2		This investigation is an interesting application of the learning in this unit. Pupils may be familiar with the idea from films but will probably be surprised at its use here. This is a good opportunity to demonstrate links to other subjects, in this case history.
Chapters 11–13 assessment on Collins Connect				
14 Revision and GCSE preparation	• Practice • Revision • GCSE-type questions	6	This chapter is going to: • Help pupils to practise and revise topics covered in their current course • Get pupils started on their GCSE course	The exercises in this chapter of the Pupil Book cover the following mathematical strands: • Algebra • Geometry and measures • Statistics • Number The material will provide excellent practice so that pupils become mathematically fluent. Encourage pupils to work through this whole chapter before their End of Year 9 tests.
Chapter 14 assessment on Collins Connect				
End of year assessment on Collins Connect				

2-year scheme of work

The following scheme of work provides suggestions for teaching Pupil Book 3.1 as part of a 2-year Key Stage 3 course.

Please note that you can recombine the test questions provided on Collins Connect to create new tests if your frequency of assessment differs from that below, or if you wish to combine content from different chapters in your own half-term tests.

This scheme of work is provided in editable Word and Excel format on the CD-ROM accompanying this Teacher Pack.

Chapter	Lesson	No. of hours	Learning objective	Comments/ suggestions
Half-term / Term 1				
1 Percentages	1.1 Simple interest	1	• To understand what simple interest is • To solve problems involving simple interest	Although pupils have met percentages before there are some important and quite challenging concepts in this chapter for pupils working at this level. The ideas of percentages as a multiplier and the use of multiplicative reasoning are very important to pupils' confidence and fluency when working with percentages. So, while you may be able to leave out some of the earlier questions in each exercise, be careful not to leave out too much or move on too fast.
	1.2 Percentage increases and decreases	1	• To calculate the result of a percentage increase or decrease • To choose the most appropriate method to calculate a percentage change	
	1.3 Calculating the original value		• Given the result of a percentage change, to calculate the original value	
	1.4 Using percentages	1	• To revise the links within fractions, decimals and percentages • To choose the correct calculation to work out a percentage	
	Challenge – The Royal Albert Hall	1		This challenge gives pupils the opportunity to extend their learning to a real-life context. All the information pupils will need is provided in the Pupil Book but it is quite complex. Pupils working at this level may find it difficult to access the information they need. This is representative of how they are likely to find information presented in real life. Pupils will need to read the questions very carefully to decide what information they need and what mathematical skills to use in each case.
2 Equations and formulae	2.1 Multiplying out brackets	1	• To multiply out brackets	Much of this chapter will be unfamiliar to pupils. However, some pupils may be familiar with expanding brackets. Check that all pupils can expand brackets fluently before moving on to the rest of the chapter. If pupils grasp the concepts quickly they can move on to the more challenging questions that are towards the end of each exercise in the Pupil Book.
	2.2 Factorising algebraic expressions	1	• To factorise expressions	
	2.3 Equations with brackets	1	• To solve equations with one or more sets of brackets	
	2.4 Equations with fractions	1	• To solve equations involving fractions	
	2.5 Formulae	1	• To practise using formulae	

	Financial skills – Wedding day	1		This financial skills activity gives pupils the opportunity to apply the skills they have learned in the chapter to a practical situation that many pupils may experience in the future. The cost formula used is often encountered in GCSE exams, so it is important for pupils have a good grasp of this.
3 Polygons	3.1 Polygons	1	• To know the names of polygons • To know the difference between an irregular polygon and a regular • To work out the sum of the interior angles of a polygon	Lesson 3.1 should be familiar material for pupils. Check pupils' knowledge by giving them some questions. If all pupils can answer them and you are satisfied that everyone in the class understands the material, then move on to Lesson 3.2.
	3.2 Angles in polygons			
	3.3 Interior angles of regular polygons	1	• To work out the sizes of the interior angles in regular polygons	
	Activity – Regular polygons and tessellations	1		This activity is designed to give pupils the opportunity to apply what they have learnt about the characteristics of polygons to tessellations. Pupils will need to apply what they know about angles in polygons. Tessellations were not part of this chapter but pupils should have met the concept before.
Holidays				
Half-term / Term 2				
4 Using data	4.1 Scatter graphs and correlation	1	• To infer a correlation from two related scatter graphs	Much of the material in the lessons of this chapter will be new to pupils. Lesson 4.3 and Lesson 4.4 could, however, be combined. Make certain that pupils have a good grasp of correlation and time series before moving on.
	4.2 Interpreting graphs and diagrams	1	• To use and interpret a variety of graphs and diagrams	
	4.3 Two-way tables 4.4 Comparing two or more sets of data	1	• To interpret a variety of two-way tables • To compare two sets of data from statistical tables and diagrams	
	4.5 Statistical investigation s	1	• To plan a statistical investigation	
	Challenge – Rainforest deforestation	1		Talk to pupils about deforestation and the fact that for years, the big rainforests of the world have been reduced and chopped down, while the country or countries concerned benefit from the cleared land and the revenue from the wood obtained from the trees. Note that this challenge does not intend to make any judgement values of the country or countries concerned. Instead, it has been devised to allow pupils to find what the statistics may suggest; in other words, that economic growth can affect the amount of deforestation.
Chapters 1–4 assessment on Collins Connect				

5 Circles	5.1 The formula for the circumference of a circle	1	• To calculate the circumference of a circle	Pupils will have met the formulae for area and circumference of a circle in Year 8. Check pupils' understanding by giving them some examples and go through the more formal explanation for area at the beginning of Lesson 5.2. If pupils are confident and fluent, move directly to Lesson 5.3.
	5.2 The formula for the area of a circle	1	• To calculate the area of a circle	
	5.3 Mixed problems	1	• To solve problems involving the circumference and area of a circle	
	Financial skills – Athletics stadium	1		This activity is designed to give pupils the opportunity to apply their knowledge to a multi-step real-life problem. The context is familiar, but the activity is presented in a slightly more complex way than pupils may be used to. All the information required to answer the questions is in the text, but pupils will need to read and then think carefully about how they access the information. Remind them to highlight the key information they will need. Tell pupils that they will need to combine their skills not only from this chapter, but also from different areas of mathematics, for example, number.
6 Enlargements	6.1 Scale factors and enlargements	1	• To use a scale factor to show an enlargement	If pupils in the class grasp concepts quickly, then it will be possible for you to combine Lesson 6.1 and Lesson 6.2. Encourage more able pupils to move straight to the more challenging questions towards and at the end of each exercise in this chapter.
	6.2 The centre of enlargement	2	• To enlarge a shape about a centre of enlargement	
	6.3 Enlargements on grids	1	• To enlarge a shape on a coordinate grid	
	Problem solving – Photographs	1		This problem-solving activity consolidates topics previously covered on extracting data, area and ratio.
7 Fractions	7.1 Adding and subtracting fractions	1	• To add or subtract any two fractions	The material in Lesson 7.1 should be familiar to pupils but pupils working at this level are likely to need reinforcing of the work. Check pupils' understanding by working through some examples. Only if appropriate, move on to Lesson 7.2.
	7.2 Multiplying fractions	1	• To multiply two fractions	
	7.3 Dividing fractions	1	• To divide one fraction by another	
	Problem solving – The 2016 Olympic Games in Rio	1		Pupils apply their understanding of fractions to a topical but more complex problem. Pupils need to work methodically to identify all the information required to answer the questions. Remind them to highlight the key information they will need. Pupils will also need to combine their understanding across fractions, decimals and percentages as well as their understanding of averages.
Chapters 5–7 assessment on Collins Connect				

Holidays				
Half-term / Term 3				
8 Algebra	8.1 Expanding brackets	1	• To multiply out brackets with a variable outside them	Much of the work in this chapter will be new to pupils, although they will know certain concepts, which are expanded on from Chapter 2. You could fast-track those pupils who grasp the material quickly to the more challenging questions at the end of each exercise in the Pupil Book.
	8.2 Factorising algebraic expressions	1	• To factorise expressions	
	8.3 Expand and simplify	1	• To expand expressions with two brackets and simplify them	
	Challenge – California gold	1		This challenge activity requires pupils to apply their learning in an unfamiliar context. Introduce it with some recent examples of treasure trove finds from the internet and get pupils to research the current price of gold per gram.
9 Decimal numbers	9.1 Multiplication of decimals 9.2 Powers of 10	1	• To practise multiplying decimal numbers • To understand and work with both positive and negative powers of ten	Pupils should have met most of the material in this chapter before. However, this material may challenge some pupils. It is important to remember that lack of confidence and fluency with basic number skills can be a significant barrier to further learning for pupils working at this level. However, if you feel your pupils are able to move on faster, you could combine Lesson 9.1 and Lesson 9.4, and then Lesson 9.2 and Lesson 9.3, by choosing key questions in each pair of lessons. Then move on to Lesson 9.5.
	9.3 Rounding suitably 9.4 Dividing decimals	1	• To round numbers, where necessary, to an appropriate or suitable degree of accuracy • To confirm ability to divide with decimals	
	9.5 Solving problems	1	• To solve real-life problems involving multiplication or division	
	Mathematical reasoning – Paper	1		This activity uses the context of paper, with which pupils may be very familiar. All the information pupils need is provided in the text in the Pupil Book, but it is quite complex. Pupils will need to read the questions very carefully to decide on the information that they will need and what mathematical skills to use in each case. The questions move freely between fractions and decimals. This is something that pupils need to be comfortable with, which also develops their conceptual understanding of fractions and decimals being ways of expressing parts of a whole.
10 Surface area and volume of 3D shapes	10.1 Surface areas of cubes and cuboids	2	• To work out the surface areas of cubes and cuboids	Pupils should be familiar with many of the concepts in this chapter. Check pupils' understanding by giving them different examples to see if they have any problems finding the answers. Once you are happy that pupils are confident, move on to the MR questions towards the end of each exercise, and the investigation and problem-solving activities at the end of each exercise.
	10.2 Volume formulae for cubes and cuboids	1	• To use a simple formulae to work out the volume of a cube or cuboid • To work out the capacity of a cube or cuboid	
	10.3 Volume of triangular prisms	1	• To work out the volume of a triangular prism	

	Investigation – A cube investigation	2		Pupils apply their understanding of area to a more complex problem. Pupils need to work methodically and be able to explain their solutions. This is a good transferable skills objective to share with pupils when doing this investigation. Ask pupils to share not only their solutions but also *how* they approached working on the problem.
Chapters 8–10 assessment on Collins Connect				
11 Solving equations graphically	11.1 Graphs from equations in the form $y = mx + c$ 11.2 Problems involving straight-line graphs	1	• To draw any linear graph from any linear equation • To solve a linear equation from a graph • To draw graphs to solve some problems	Pupils may be familiar with the material in the first two lessons of this chapter. Check pupils' understanding by giving them some well-targeted questions about $y = mx + c$. If they are confident, you may want to combine Lesson 11.1 and Lesson 11.2 using the MR and PS questions, and the end of lesson activities.
	11.3 Solving simple quadratic equations by drawing graphs	1	• To solve a simple quadratic equation by drawing a graph	
	11.4 Problems involving quadratic graphs	1	• To solve problems that use quadratic graphs	
	Problem solving – Squirrels	2		Pupils often ask why they have to do mathematics that is not familiar to them. Say that using graphs to monitor wildlife is a good example of how mathematics can be used to benefit society.
Half-term / Term 4				
12 Distance, speed and time	12.1 Distance 12.2 Speed	2	• To work out the distance travelled in a certain time at a given speed • To use and interpret distance–time graphs • To work out the speed of an object, given the distance travelled and the time taken	The material in all three lessons of this chapter will be new to pupils. However, if you feel that your pupils are confident with the different units for speed, distance and time, then you could combine the three lessons into one lesson. You could do this by using the speed, distance, time triangle.
	12.3 Time	1	• To work out the time an object will take on a journey, given its speed and the distance travelled	
	Financial skills – Shopping at the market	1		This financial skills exercise will help pupils to make their learning relevant by applying it to a real-life situation.
13 Similar triangles	13.1 Similar triangles	2	• To understand what similar triangles are	This chapter is new material and in many cases quite complex. Choose examples carefully to support or challenge pupils.
	13.2 A summary of similar triangles	1	• To understand what similar triangles are	

	13.3 Using triangles to solve problems	1	• To understand that triangles can be used to solve some real-life problems	
	Investigation – Barnes Wallis and the bouncing bomb	1		This investigation is an interesting application of the learning in this unit. Pupils may be familiar with the idea from films but will probably be surprised at its use here. This is a good opportunity to demonstrate links to other subjects, in this case history.
14 Revision and GCSE preparation	• Practice • Revision • GCSE-type questions	6	This chapter is going to: • Help pupils to practise and revise topics covered in their current course • Get pupils started on their GCSE course	The exercises in this chapter of the Pupil Book cover the following mathematical strands: • Algebra • Geometry and measures • Statistics • Number The material will provide excellent practice so that pupils become mathematically fluent. Encourage pupils to work through this whole chapter before their End of Year 9 tests.
Chapters 11–14 assessment on Collins Connect				
End of year assessment on Collins Connect				

Notes

Notes

Notes

Notes

Notes

Notes